MatThink ™

to optimize materials

The Exemplary Worker Book Series

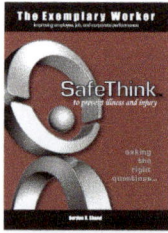

SafeThink™ ...to prevent illness and injury

SafeThink is a structured critical thinking strategy you can use to identify, predict, and control hazardous situations before, during, and after completing work. This cognitive-based safety strategy can be used on the fly, at work, at home, at play, and while driving. *SafeThink* also provides strategies for you to remain focused on your tasks.

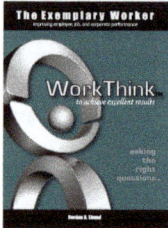

WorkThink™ ...to achieve excellent results

WorkThink is a thinking strategy you can use to achieve quality results with the least amount of effort. It usually takes little extra effort to do quality work instead of inferior work. *WorkThink* also emphasizes understanding the expectations of your supervisor, team leader, and customers so that you can achieve the excellent results they expect.

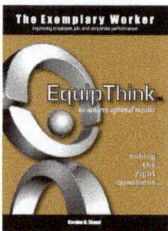

EquipThink™ ...to achieve optimal results

EquipThink is a thinking strategy for you to use tools, mobile equipment, and stationary equipment effectively and efficiently. The goals are for you to achieve the desired results with minimal stress on equipment, to conserve energy, and to extend equipment life. The input–process–output thinking strategy, in conjunction with identifying critical variables, is used to achieve optimal results.

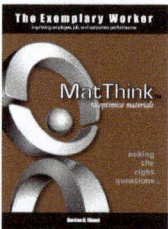

MatThink™ ...to optimize materials

MatThink is a thinking strategy you can use to make the most effective use of materials. The thinking strategy applies to recovering, processing, modifying, applying, transporting, and storing materials. Because equipment and materials are usually closely related, the input–process–output thinking strategy, in conjunction with identifying critical variables, is used to optimize material recovery and use.

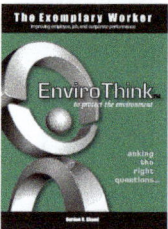

EnviroThink™ ...to protect the environment

Both industry and individuals have a responsibility to protect the environment. *EnviroThink* is a critical thinking strategy you can use to identify and respond to environmental issues for any job position that you might hold. *EnviroThink* helps you think through your work by asking yourself specific questions relating to environmental issues important to organizations.

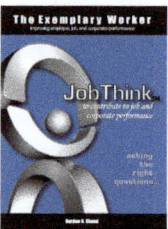

JobThink™ ...to contribute to job and corporate performance

Exemplary workers understand what is important to their organizations. They know the issues critical to business success and where to focus their efforts. *JobThink* addresses the critical thinking strategies you can use to identify what is important for job and corporate performance.

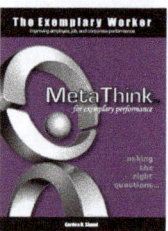

MetaThink™ ...for exemplary performance

MetaThink applies some of the thinking strategies addressed in previous books in different ways and also addresses new thinking strategies useful for the workplace. You can use these thinking strategies, along with the detailed thinking strategies addressed in other books of this series, to achieve exemplary performance.

The Exemplary Worker Book Series

"Rarely can workers from any sector access self-paced instructional materials that are easy-to-use, step-by-step guides to workplace learning. *The Exemplary Worker* book set is an exception. These books offer a good breadth of learning for workers in contexts ranging from: exemplary performance; job and corporate performance; results optimization; and work excellence. With meticulous organization, these essential training references are helpful guides for workers seeking to improve their performance. With prefaces designed to help trainers/instructors assist workplace learners, these books use critical thinking strategies that identify what matters to workers and supervisors considering people, equipment, materials, environments, and organization in concert."

—**Eugene G. Kowch, Ph.D.**, Leading Complex and Adaptive Learning Systems/Organizations, University of Calgary, Canada

"The power of thinking in determining our safety, health, and welfare is obvious, but how to manage such cognition or self-talk for injury prevention, self-motivation, and self-improvement is not so obvious. Answers are provided in this action-focused series of self-help books on *The Exemplary Worker* by Gordon D. Shand. He offers much practical information for leadership, safety, and well-being. Each of these books provides critical and structured thinking strategies for optimizing performance on several fronts, from improving safety and productivity in the workplace to actively caring as a teacher, parent, or friend."

—**E. Scott Geller, Ph.D.**, author of The Psychology of Safety Handbook; Alumni Distinguished Professor, Virginia Tech; Senior Partner, Safety Performance Solutions

"These are very practical books. I, myself, have been interested in the fundamental processes of human thinking. For creativity there is Lateral Thinking. For exploration there is the parallel thinking of the Six Thinking Hats. For perception there is the CoRT school programme. *The Exemplary Worker* series of books provide frameworks for focused thinking about specific situations. The frameworks guide the thinker to deal with the situation instead of messing about. That is why the books are so practical."

—**Dr. Edward de Bono**, Author of Lateral Thinking and Six Thinking Hats and creator of CoRT

The Exemplary Worker Book Series

MatThink™

to optimize materials

Gordon D. Shand

HDC Human Development Consultants Ltd.
PO Box 4710, Edmonton, AB, Canada T6E 5G5
www.hdc.ca
www.safethink.ca

MatThink™

Library and Archives Canada Cataloguing in Publication
Shand, Gordon D.
    MatThink to optimize materials / Gordon D. Shand.
(The exemplary worker book series)
ISBN 978-1-55338-055-9
    1. Materials. 2. Critical thinking. 3. Employees--Training of.
3. Quality of products. I. HDC Human Development Consultants
II. Title. III. Series: Exemplary worker
TA403.6.S52 2014   658.7   C2014-902766-9

Published by HDC Human Development Consultants Ltd.

Published in Canada

HDC Human Development Consultants Ltd.

Website: www.hdc.ca
E-mail: hdc@hdc.ca
Phone: (780) 463-3909

Acknowledgements

Developing *The Exemplary Worker* book series has been challenging and rewarding. I am certainly grateful for all the help I have received to produce quality products. Over one hundred people have contributed to the quality of the content and presentation.

Generally, I developed the first draft of the books working on evenings and weekends. I would blitz the first draft for a book—I produced the draft in a month to three months. During those times, my family's gracious support allowed me to concentrate on the task and to dialogue with them about the concepts. Once a first draft was produced, consultants in my firm carried out several edits as time allowed. HDC's Production Department developed illustrations and formats to produce a book ready for validation by industry. Because the people from industry volunteered their time and some validations were conducted in sequence, the validation process for each book took up to six months or more.

Many staff contributed to the development process. I would like to acknowledge those consultants who struggled to gather relevant content when working with customers—they gave cause to identify the thinking strategies used by exemplary workers and to develop the training for HDC consultants. Many thanks to the consultants who worked so diligently with me to produce the books. They were adamant in adhering to our standards for quality, even when I was burned out and wanted to put closure to a topic. Thanks to Janelle Beblow, Art Deane, Alice Graham, Jean MacGregor, and Bruno Schoenfelder for the wonderful edits and feedback. Thanks to Phil Jenkins, Kris Vasey, and Denise Hodgins for developing the illustrations, formatting the documents, and creating the book covers. Thanks to Maria Peck for coordinating the validations and field tests and proofing text. Their personal support, commitment to quality, and attention to detail are greatly appreciated.

I have been exceptionally fortunate to work with so many wonderful people from industry. They have been great mentors—they have made many contributions to my personal growth. A special thanks to nearly a hundred people who have volunteered their time to validate and field test the strategies.

Who is *The Exemplary Worker* series for?

The Exemplary Worker series benefits:

- **Individuals** who want to have outstanding performance
- **Apprentices and students** who want to work safely and effectively
- **Supervisors** who want staff to be more effective
- **Trainers** who want to contribute to improved corporate, job, and employee performance
- **Trades and technology instructors** who want their apprentices and students to work safely and effectively
- **Instructional designers** who want to ensure that training is relevant, useful, and practical
- **HR managers** who want to improve the development and retention of exemplary workers
- **Operations staff** who want to optimize production and minimize losses

Contents

MatThink™

Table of Contents (continued)

Preface

In addition to being skilled, exemplary workers use a broad range of *critical thinking strategies* to maintain outstanding performance. Exemplary workers know what is important to their jobs and organizations—they put their efforts in the right places by doing the most important things, doing them effectively, and doing them efficiently. Because they know what is important to the job and the organization, they effectively coordinate their actions with others and make decisions in the best interest of their organizations. Knowledge and thinking skills empower workers to achieve exemplary performance, be flexible as workplaces continue to evolve, and provide leadership within the workplace.

Exemplary performance can have many benefits for you, the line worker, lead operator, foreman, or supervisor, including:
- increased job satisfaction
- being recognized by your peers and supervisors as an effective employee
- increased potential for keeping your job during slow economic times
- increased potential for receiving salary/wage increases or bonuses
- increased opportunity for new or different work assignments
- increased potential for promotion

Each of the seven books in *The Exemplary Worker* series focuses on one of five domains (**PEMEO**):
- **P**eople
- **E**quipment
- **M**aterials
- **E**nvironment
- **O**rganization

Loss and/or optimization (LO) are the main themes for the domains, creating the word **LO-PEMEO™**. LO-PEMEO stands for Loss and Optimization of People, Equipment, Materials, Environment, and Organization. As an example: **L**oss to **P**eople is illness and injury; **O**ptimizing **P**eoples' performance is working effectively and efficiently; **L**oss to **E**quipment is damage and shortened operating life; and **O**ptimizing **E**quipment is using equipment effectively and efficiently. The books place a strong emphasis on using **thinking strategies** and **asking quality questions**—the goals are to minimize losses and optimize performance of PEMEO.

The series of books addresses both loss and optimization of each domain. We recommend that you complete each of the first six books in the sequence. However, the books can be studied in any order without difficulty. The last book in the sequence, *MetaThink*, should be read last. *MetaThink* applies some of the thinking strategies addressed in previous books but in different ways and also addresses new thinking strategies useful for the workplace.

Introduction to *The Exemplary Worker* Series

Over the last twenty-five years, the process of discovering *what's important* for exemplary worker performance has gone full circle. The process began for me when I interviewed exemplary workers to identify relevant training content. My premise was that exemplary workers know what is important for people to do their jobs effectively. Over time, it became apparent to me that one of the reasons exemplary workers perform so well is that they use a set of generic thinking strategies. After starting a consulting firm to design and develop training, I developed a comprehensive internal training program for our consultants and technical writers who develop training programs. The training focused on using generic thinking strategies and critical questions to identify training content that helps workers perform effectively. With a lot of support, I have revised our consultant training program and made it available to the public for people to learn and refine their personal thinking strategies to be exemplary workers.

The Exemplary Worker books are presented as a series. The same concepts underlie all seven books. For example, a safety incident may cause harm to a person and result in other losses—work may be suspended, equipment and materials damaged, the environment harmed. The organization could also experience unpredicted costs and have its reputation harmed. This introduction provides a framework and the key concepts that apply to the series. The discovery process and happenstances that led to the development of *The Exemplary Worker* series are explained to provide a setting and context to give meaning to the underlying concepts.

The Discovery Process

For me, the real discovery process began in 1985 when I founded the consulting firm HDC Human Development Consultants Ltd. (HDC) to design and develop customized technical training programs. I believed that it was possible to develop quality training for any industry without having an in-depth understanding of the organization, its technology, or the tasks that its people perform. The premise was that a well-thought-out instructional design and development process combined with effective consulting skills would be sufficient.

As founder of the company, I felt that I was successful in providing leadership to identify training content important to my customers—customers often asked me to do additional work. If I could do the work well, then certainly others in the firm could as well and, for some deliverables, do better.

The Plan

The plan was that I would work with customers to develop the outline of the training program (curriculum) and identify critical content for the program. The training program would be documented in one of three ways:
• a list of specific courses
• a list of general training objectives
• a competency-based training profile

Competency-Based Training Profile

The following illustration is a *partial* example of a competency-based training profile. The profile is a visual presentation of the competencies (tasks and support knowledge) that specific work groups require to do their work safely and effectively.

MatThink™

ORIENTATION	Complete Company Orientation	Describe Roles and Responsibilities	Identify Local Structures and Facilities	Describe and Use Communication Systems	Identify Customers and their Expectations
SAFETY	Describe and Use Personal Safety Equipment	Review Safety Handbook	Complete First Aid Training	Decribe and Operate Personal Gas Monitors	Describe Codes of Practice
ENVIRONMENT	Describe Environmental Responsibilities	Describe and Store Hazardous Wastes	Describe and Monitor Gas Emissions	Take Waste Water Samples	Describe and Participate in Spill Response Exercises
GENERAL KNOWLEDGE AND SKILLS	Describe Flammable Gas Measurements	Use Portable Multi-Gas Monitor	Describe Reciprocating Compressors	Prepare Maintenance Requests	
ROUTINE TASKS	Carry out Routine Equipment Checks	Change Process Filters	Describe and Change Corrosion Coupons	Monitor and Adjust Inhibitor Injection	Perform Housekeeping
SITE-SPECIFIC KNOWLEDGE AND TASKS	Describe Remore Process	Start and Adjust Remore Process	Describe and Change Remore Output Parameters	Perform Emergency Shutdown of Remore Process	Shut down Remore Process for Maintenance

Critical content for each competency is a list of the key issues a buddy or supervisor would emphasize when coaching the trainee. The end product is a *scope document* listing the key issues and ensuring continuity between competencies—no overlaps or gaps in content. As an example of a scope document, here is a partial list of key issues for the competency *Purge Piping and Station Systems:*

- replacing one medium with another to prevent combustible or toxic condition
- important to prevent:
  - people being exposed to toxic gases
  - possibility of a fire
- piping should only be purged after system has been opened and exposed to a foreign substance
- stations purged in preparation for startup
- some stations have automatic purging for specific piping and equipment
- automatic purging sequence must be checked
- always purge in direction gas migrates (up or down)
- criteria for length of time to purge include volume, pressure, and amount of connected equipment

In a profiling workshop, I used a brainstorming technique with four to sixteen of the customer's employees to identify competencies and critical content. The workshops were mentally demanding. On the one hand, I was concerned that the scope of training and performance requirements be limited and only address competencies and content that were considered important to the workers, their supervisors, and the organization. On the other hand, I was concerned that critical issues affecting people and the business were not overlooked. During these workshop sessions, I was constantly searching for relevant, useful, and practical content. What do the workers do? Is there a special way of doing the task? How do they know they are doing a good job? What can go wrong? How can the equipment be damaged or its life shortened? What do you mean by product quality? What about safety and the environment? Does the organization have special policies and ways of doing business? What is important and to whom or what? What questions should I be asking the group? I did not have a clear set of criteria or a structured thinking process that I could use to provide leadership in identifying training content that was important to the worker and the supervisor.

Working with Subject Matter Experts (SMEs)

I certainly believed that asking quality questions was more important than providing content. Answers to the questions could be provided by the customer's experienced employees. The term *subject matter experts* (*SME*s) is often used to refer to the organization's staff who provide content to training consultants and technical writers. Unfortunately, some SMEs, having in-depth knowledge of the tasks, technology, and the organization, had difficulties identifying content important for training. These SMEs expected consultants to provide leadership to identify relevant content. I soon discovered that my consultants often had difficulties in providing leadership to SMEs trying to identify content that was relevant, practical, and useful. When reviewing the first draft of training modules, information that would help trainees do their jobs more effectively, efficiently, and safely would often be missing. Nor would the supervisor's concerns always be addressed. Sometimes, information would be included that was of little value in helping workers do their jobs well and making decisions in the best interest of their organizations. When consultants asked me for direction as to the types of content that were relevant for training, I could not provide a comprehensive explanation. If the company was going to be successful in the future, I needed to find ways to define content that was relevant, practical, and useful—content that contributed to employee, job, and corporate performance.

Customer feedback gave me reason to believe that I was providing adequate leadership to identify relevant content; that I was asking quality questions. The truth of the matter was I did not have a formal list of types of question I should ask. In many ways, I was relying on intuition to ask the right questions. I needed to find a way to articulate a content gathering strategy that consultants could use with a variety of customers in different lines of business, different technologies, different hiring practices and performance expectations, and different ways of conducting business. I needed to find a way to identify the specific types of question consultants could ask SMEs to identify important training content—content that would help workers perform their jobs safely and effectively and contribute to meeting corporate objectives.

To help our training consultants and technical writers gain a better understanding of our customers, their businesses, goals, and concerns, I took consultants along to the competency-based profiling sessions. Listening to the group discussions and individual insights about the work and the business always provided learning beyond the information recorded in the program outline and scope document. This learning should be valuable when working with SMEs to identify detailed content for the training resources. Having this preliminary knowledge about the customer seemed to help some consultants be better at identifying relevant training content, but other consultants continued to struggle. I concluded that knowledge about the customer was valuable but didn't give consultants the strategies they needed to provide leadership when working with SMEs.

The Importance of Training Content Being Relevant to the Organization, Job, and Employees

Project reviews with customers were very useful for gaining ideas on how to improve services and products. Feedback from SMEs was that HDC consultants asked more questions than anyone they had ever worked with before. On the other hand, our consultants felt that they didn't ask enough questions because relevant information had been missed. The real issue was to ask fewer questions but more *quality* questions— questions that addressed issues that were important to employees, the job, and the organization. Certainly, customers strongly indicated that identifying relevant, useful, and practical content was the most important quality concern they had regarding the development of training resources. Customers also were adamant that consultants provide direction and leadership when working with SMEs to identify relevant content.

At the close of each project, I would ask the customer what additional training might be useful for consultants to help them be more effective at identifying relevant content. Suggestions included that consultants could increase their technical knowledge, or have a better understanding about safety management systems, environment management systems, or management styles. In response to suggestions, we began providing additional internal training using off-the-shelf technical training materials when possible. The additional training helped consultants to better understand what SMEs were telling them but only resulted in marginal improvements in consultants being able to provide leadership to identify relevant content. I concluded that the knowledge is useful but not sufficient in helping consultants (and workers) to identify issues important to employee, job, and corporate performance.

To compound the problem of identifying relevant content, expectations in industry were changing from developing entry level training (do as I tell and show you and don't ask why) to exemplary level training (maximizing productivity and making quality decisions) and every level between those two extremes. These changing expectations created difficulties in determining the content and amount of detail to include in training and keeping within training development budgets. Customers were upset if training materials included content they did not want and were not willing to pay for. Customers could also be disappointed if the training did not include content that they considered important. In many ways, the concerns consultants had in understanding the customer's expectations are the same concerns an employee new to a job would have.

When I had worked with exemplary workers, I discovered that one of their strategies was to confirm expectations. So we used the same strategy and built more confirmation checks into the development process to ensure the content was what customers wanted. Unfortunately, the confirmation checks were good at confirming that the documented content was what customers wanted but did not effectively address concerns about omissions of content important to customers (e.g., safety, equipment life).

Identifying Thinking Strategies Used by Exemplary Workers

Developing internal training for consultants to effectively identify relevant, useful, and practical content proved to be very difficult. Having consultants participate in the profiling sessions to learn about the customer, developing scope documents, providing technical and organization training, and building in confirmation checks had some value but weren't sufficient in helping them to provide leadership to identify relevant training content.

The instructional systems design models I was familiar with generally placed a strong emphasis on instructional development processes and only provided marginal direction and strategies on how to provide leadership to identify content that was important to customers. Certainly, the design of instruction and the nature of the content had an effect on each other. I suspected that there were instructional designs in which generic module structures and generic types of content would work for some types of technology and associated training outcomes. It would be several more years, after we had a large inventory of customized self-instructional modules, before we were able to develop a set of generic boilerplates (list of section and sub-section titles) for specific technologies and training outcomes. These *boilerplates* provided general structures for self-instruction and listed the types of content that *could* be included (but not necessarily included) in each section. No doubt, the SMEs that I worked with had mentally created their own boilerplates to be effective when working with specific types of equipment.

My initial effort to develop training to identify relevant content proved to be fairly impractical. Fortunately, several events provided me with the fundamental concepts needed to develop strategies that consultants could use to identify relevant content.

One of HDC's customers had a very demanding supervisor who was exceptionally analytical. In fact, he was by far the most powerful analytical thinker I have met. He was also driven to prevent anything negative from happening. He would always be analyzing situations and wanted to know all the *hows* and *whys* about every aspect of the instructional design that came to mind. Once a week I would make a personal visit to address his concerns. On one of those visits, he demanded to know what type of content should be addressed in the training. He said he asked our consultant the same question and the consultant's response was that *he would write self-instruction on anything as long as we told him the content.* Obviously, the consultant was not providing leadership when working with the SME to identify training content that would help the operators perform their work safely and effectively. For me, it was confirmation that our internal training was not very effective in helping our consultants to provide leadership.

My immediate response to his demand was to give some general criteria for identifying relevant content. *Well, safety, environment, equipment life, product quality, and customer satisfaction are important. Adhering to legislation and making decisions are important, too.*

There was a long silence—a lot of mental processing was going on in his head. Finally, he nodded and said, *Good. Let's tell the consultant and the senior operator what you just said.* The bottom line for this customer was that the training we were developing would contribute to his staff doing their work effectively and safely and making good decisions.

The interaction I had with that customer was the moment of discovery for me! The three-hour drive back to the office gave me time to reflect on what had just happened. Obviously, until I was asked, I had not been able to see the forest for the trees. Ask any business person what is important to their business success and he or she would give a list of areas of concern similar to the one I gave to my customer. No doubt the business person's list would be more extensive and include additional concerns affecting productivity and controlling losses—all businesses want to get the most out of their assets, including their people. Businesses prefer to have exemplary workers, workers that contribute to business success. Certainly, the training we develop for customers must help workers be effective in doing their jobs.

Creating the LO-PEMEO Model to Identify Relevant Training

I reflected on the thinking process I was using to identify relevant content when developing training profiles and scope documents. The questions that I had been asking myself during the sessions addressed the optimization and prevention of losses primarily to People, Equipment, Materials, Environment, and the Organization as a whole. Surely, the questions would

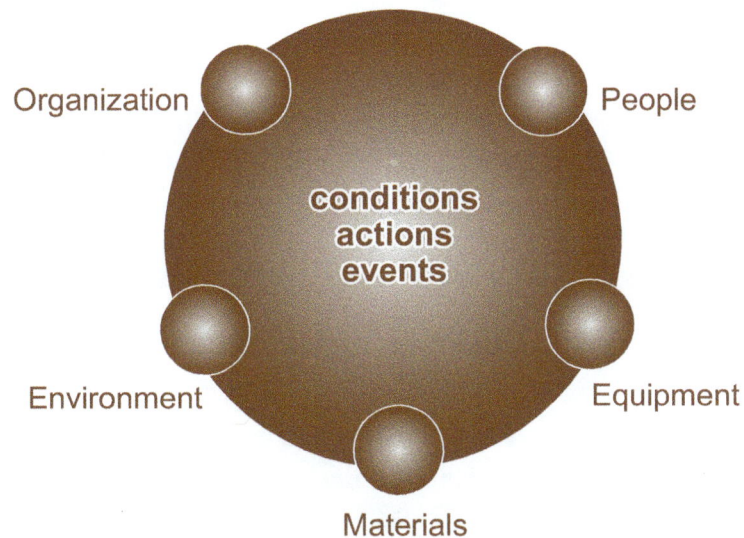

take on meaning when the work environment was considered. And one way of assessing the work environment was to consider the conditions, actions, and events within the workplace that affect PEMEO.

Most exciting for me, I could combine the concepts of optimization and controlling losses of organizational assets such as people and equipment to create a model and strategy for identifying relevant content. The LO-PEMEO model was born. Each of the five domains (people, equipment, materials, etc.) shown in the above illustration had potential for optimization and loss. An example of loss to people is illness and injury. Loss of materials when processing ore is the inefficient recovery of the desired products. Optimization of materials in construction is to use the right materials and maximize the use of the materials. The following illustration shows the combinations of loss and optimization of PEMEO.

LOSS				OPTIMIZATION	
Loss: People	LP	P	OP	Optimization:	People
Loss: Equipment	LE	E	OE	Optimization:	Equipment
Loss: Materials	LM	M	OM	Optimization:	Materials
Loss: Environment	LE	E	OE	Optimization:	Environment
Loss: Organization	LO	O	OO	Optimization:	Organization

Exemplary workers consider the potential for Loss and Optimization of each domain of PEMEO (i.e., LO-PEMEO) while they work. So LO-PEMEO was used as the framework and structure for *The Exemplary Worker* series of books. For example, loss to people (LP) is safety—the book *SafeThink* focuses on using a structured critical thinking strategy to identify and predict hazardous situations to prevent illness and injury.

Interestingly, several years later, I was introduced to a loss control model created by Frank E. Bird that used PEME as an acronym. I have always wondered if it would have saved me a lot of effort if I had known of Bird's loss control model earlier. Or would that knowledge have put in place constraints such that I would never have created the LO-PEMEO model?

While driving back to my office, I thought about how fortunate I had been over the years to work with a lot of exemplary performers, many of them my SMEs. Our customers gave us SMEs who are exemplary workers because the belief is that exemplary workers know what is important for business success and will provide training content that is relevant to corporate, job, and employee performance. When I had asked the SMEs if there were any concerns about issues such as safety, equipment, or materials, they would often look at the ceiling and ponder for a while. If they said yes, they would go on and give me further clarification. If they said no, I would continue to ask different questions. When I thought about it, the questions that I asked SMEs usually focused on concerns about LO-PEMEO. I always wondered what the SMEs were thinking when they were looking at the ceiling and pondering the answers to my questions. Eventually, I asked them. Interestingly, different SMEs from different companies and lines of business had similar concerns. For example, damage to equipment often involved shock from a sudden change in physical forces or temperature. The sources for causing damage could be people, material, or any of the other three domains. In fact, *each domain has the potential to affect the other domains.* Whether the SMEs were aware of it or not, they were mentally searching for specific workplace concerns relating to LO-PEMEO. In many ways, even at the detailed level, *the thinking strategies of exemplary workers were similar and generic.* Certainly, being aware of one's own thinking strategies contributes to planning and working effectively and helps to communicate effectively when collaborating with others and mentoring apprentices.

Linking Corporate, Job, and Employee Performance

When organizations develop standards, procedures, and training, they want to realize an improvement in corporate performance. Improving *corporate performance* is often achieved by either filling a gap in performance or by preparing the organization to move towards new goals. The following illustration lists some criteria that can be used to measure corporate performance.

PERFORMANCE REPORT

Customer Satisfaction	UP
Production	UP
Product Quality	UP
Equipment Run Time	UP
Equipment Damage	DOWN
Energy Consumption	DOWN
Material Waste	DOWN
Personal Injuries	DOWN
Maintenance Costs	DOWN
Environment Damage	DOWN
Rework Time	DOWN

At the operational or job level, the supervisor also has concerns about performance. Within his or her roles, responsibilities, and authority, the supervisor is expected to maximize productivity and minimize losses. Improved *job performance* contributes to improved corporate performance. The supervisor therefore represents the concerns and goals of the organization and must use specific resources and assets (including people) to effectively achieve the goals. The supervisor must also be able to motivate, coordinate, and assign staff to effectively carry out the work. Furthermore, worker performance affects job performance which, in turn, affects corporate performance.

Employee performance affects business results. Employees are expected to work effectively and efficiently and make good use of materials and technology. Expectations of performance are articulated to line employees both orally and in writing. In turn, employees have concerns about understanding the expectations and working safely, effectively, and efficiently to meet the expectations. The following illustration is of a person new to a job asking questions relating to corporate, job, and employee performance issues.

MatThink™

What's important to the business?

What does the team leader expect of me?

What am I supposed to do?

How am I supposed to do it?

How do I know I've done well?

How does my work affect others?

Is there a better way?

What tools and equipment are used?

Could I get hurt?

Could I injure others?

Could I damage the equipment?

Does this product affect the environment?

How much waste is acceptable?

How can I prevent…?

Will the customer be satisfied?

What should I do if …?

What would happen if …?

Do I have the authority to take action?

What action?

Whom should I inform?

What does …?

How does …?

What caused …?

What is the reason?

What are the consequences for …?

What questions should I be asking?

What answers do I need?

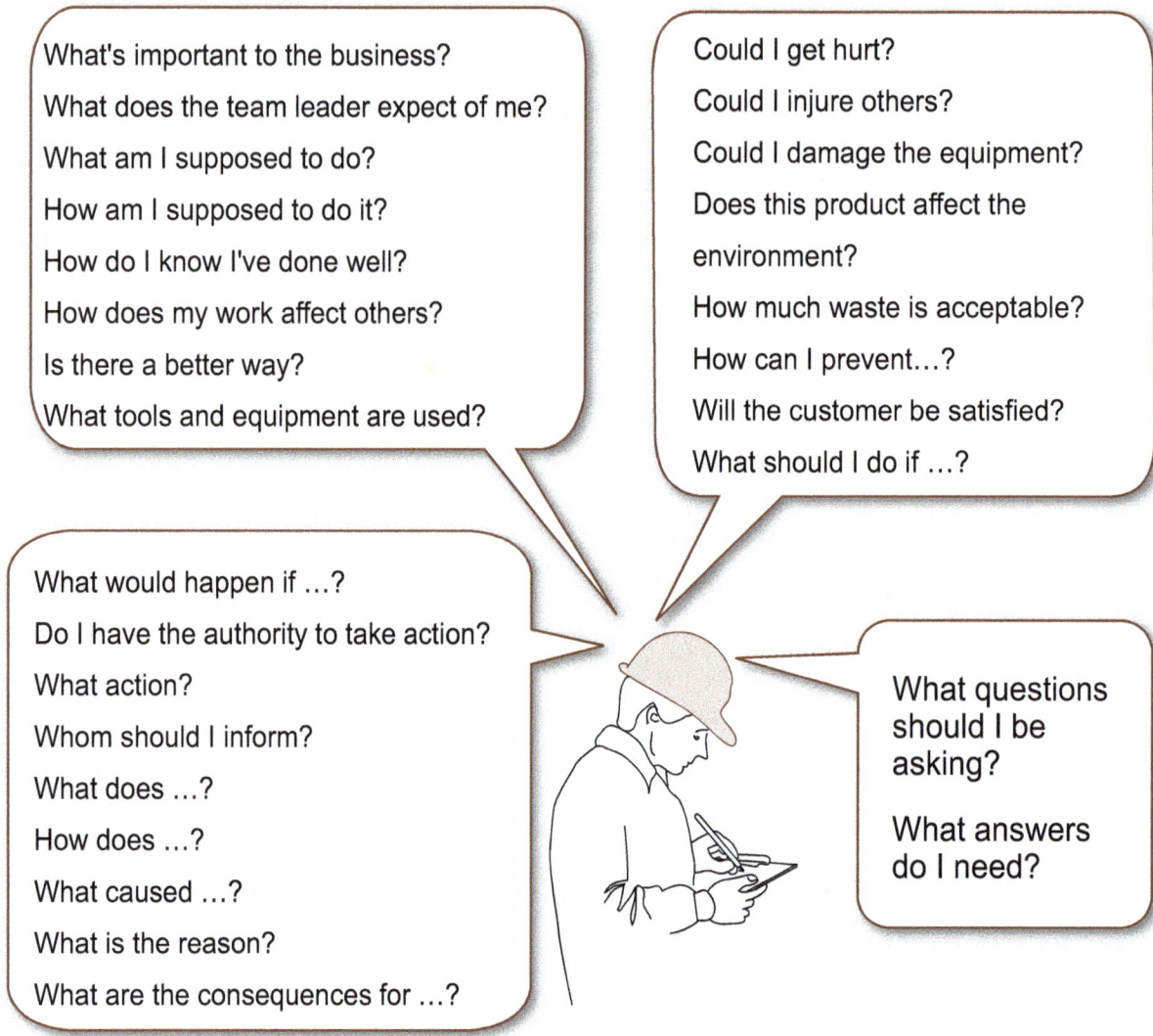

Many of the questions are generated by the LO-PEMEO strategy and focus on performance:
- What is important?
- What are the issues?
- What questions should I ask?

The person new to the job needs answers to the questions in the illustration to quickly learn to do that job effectively and efficiently. Interestingly, two employees with similar experiences and skills who are new to a job can perform quite differently. One employee will be uncertain about the work and become stressed if work conditions change. The other employee will initiate actions and make good work-related decisions for the organization within a few weeks. One of the factors that makes the difference in performance between the two employees is the knowledge about what is important to job and corporate performance. Understanding *what is important* provides criteria for focusing one's efforts and for making decisions. LO-PEMEO is a good start in identifying what is important to the organization. Although many of the issues identified by LO-PEMEO are generic, each organization has its own business strategies, resources, and priorities. As such, each organization could place a different emphasis on each issue identified by LO-PEMEO. And that's why asking the *right* questions is so valuable. Questions focus on key issues; the answers to the questions are unique to the organization, workplace, and specific circumstances. *The Exemplary Worker* series provides many of the questions that workers need to ask of themselves and of others to achieve exemplary performance.

Understanding Organizations for Exemplary Worker Performance

Exemplary workers understand what is important to the organization so that they put their efforts in the right places, do the right things, and make good decisions in the best interests of their organizations. For workers to have exemplary performance, they need to have an understanding of organizations in general, and a specific understanding of their own organization. Training and performance consultants also need to have a general understanding of organizations to be effective at developing customized training—training that is relevant, useful, practical, and reflects the organization for what it is. There is a lot of literature on organizations but most of it is more complex than training consultants need. Generally, the literature does not directly address issues important to designing and developing customized training for industry.

So, what issues are important? For consultants at HDC (and exemplary workers in other organizations) to be effective, they must be able to identify and understand organizational issues from different points of view. Imagine a roomful of statues facing in different directions. The room has many doors, each opened by a different work group or discipline. Each doorway has a different view of the statues.

For consultants to get a broader understanding of the organization, they need to view the statues from different doors. Ideally, consultants would walk around the statues to get many different points of view. The consultant must be prepared to consider different points of view within a specific organization to be effective at understanding the organization and identifying issues important to employee, job, and corporate performance.

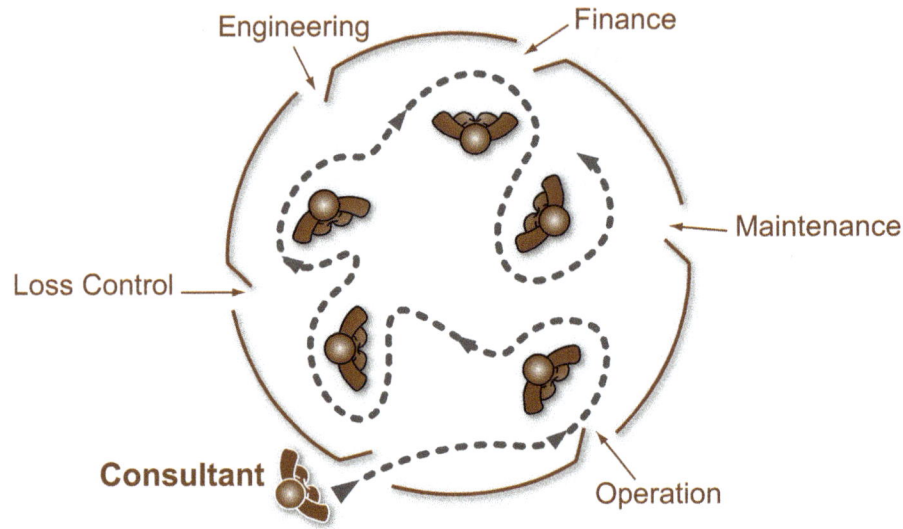

Both exemplary workers and training/performance consultants benefit from an understanding of relationships between business resources, organizational structure, business strategies, corporate objectives, and performance standards. Exemplary workers gain an understanding as to how their line of work fits into the organization as a whole. In doing so, they appreciate how their work affects others and they potentially make better use of organizational resources. This understanding about organizations also helps training consultants and technical writers to be more effective at designing and developing training that is customized, reflects the business, and has excellent value for the customer.

The approach I take with consultants to learn about organizations is to pretend to build a new business. Would the line of business be a service or a product? What is the mission? If the business is a service, then performing tasks is the main way to generate revenue and tools/equipment provide support for carrying out the work. If the line of business is to use technology to make products, then the technology dictates many of the tasks that workers must do. Having resources to achieve specific results is essential but not sufficient for business success. The resources must also be managed effectively. The following illustration identifies some key constituents of a business.

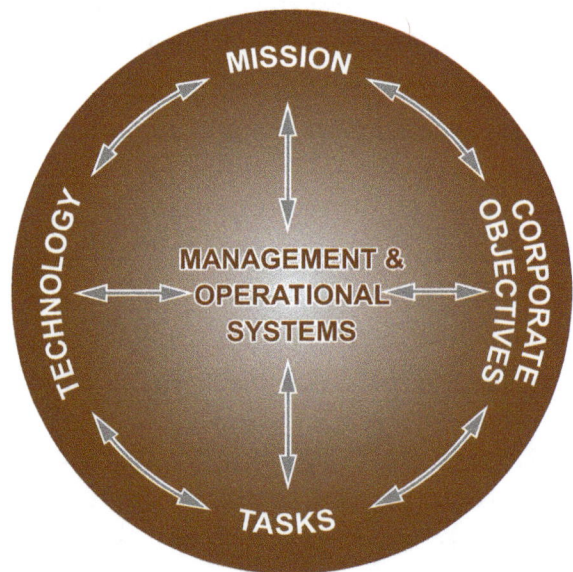

The book *JobThink* uses the previous model to provide a practical way for workers to understand organizations. This understanding helps workers to effectively focus their efforts and make decisions in the best interests of their organizations.

Of particular interest are the *corporate objectives*. Corporate objectives provide direction for using technology, performing tasks, and coordinating work to effectively achieve the corporate mission. The following table lists areas of concern, common to many organizations, for which corporate objectives may be developed.

Areas addressed by Corporate Objectives

- safety
- environment
- legislation
- equipment reliability and life
- equipment optimization
- energy use
- quality
- waste control
- loss control
- cost control
- customer satisfaction
- public image
- public disruption
- reputation
- communication
- teamwork

For a specific organization, a list of corporate objectives can be generated by expanding the organization's strategic business objectives or by using LO-PEMEO. Some companies issue strategic business objectives to provide direction to employees as to where to put their energy and focus for business success. Strategic business objectives identify what the organization must do well to be successful. For example, leaders in an organization may believe that it is essential for business success to have reliable service and satisfied customers. Organizations may identify five to eight strategic objectives. Within a department, the list of objectives (or goals) may be expanded in more detail to address issues specific to the department's mandate.

The expanded list of corporate objectives can also be generated using LO-PEMEO— each of the items in the above table relates to one or more of the LO-PEMEO domains.

Corporate objectives are fundamental to exemplary performance because they define what is *important* to the organization, the job, and workers. Corporate objectives provide a ***formal link*** between organizational goals and worker performance. Workers can use corporate objectives as criteria for working effectively and efficiently and for making decisions in the best interest of their organizations. Training consultants and technical writers can use corporate objectives to identify relevant, useful, and practical training content. Refer to my book, *Interviewing to Gather Relevant Content for Training* for:
- information about applying critical thinking skills to identify relevant content for training
- an interviewing process that consultants and technical writers can use to interview SMEs to gather relevant content

Developing Training to Identify What is Important to Employee, Job, and Corporate Performance

With the LO-PEMEO and business models, I could now develop training for consultants to provide leadership to identify relevant content. The LO-PEMEO model was the most practical approach to use to structure the training because it relates directly to work and job issues. The organizational model can be integrated into the training on loss and optimization of organization, LO-O. For the training on these models to be useful, the training needs to be flexible and apply to a broad range of work, technology, and organizations. The training must also provide strategies for people to think through their work. That is what exemplary workers do—they think through their work. And, the thinking processes are generic so they apply to all types of industries, work environments, and jobs.

All of the training to identify relevant content is founded on using thinking strategies. An emphasis is placed on *concepts* and *generalities* to maintain a broad application of the thinking strategies. Furthermore, the thinking involves asking questions relating to LO-PEMEO. Asking questions is important to maintaining the broad application of the thinking strategies and helping people remain mentally engaged. Asking the *right* questions is often more important than finding the answers, because if the right questions are asked, answers can usually be found— answers that contribute to exemplary employee, job, and corporate performance.

Over several years, I developed training for all the combinations of LO-PEMEO. I also expanded the training to include consulting processes and a performance and training model to design, develop, and implement competency-based training and performance management systems. I was very fortunate to have excellent support from staff to edit and refine the training. HDC staff made important contributions to the training content and presentation. And, after the training resources were in use, we refined them further.

Developing *The Exemplary Worker* Series

After the HDC consultants' training resources had been used for ten years, I decided to go full circle and modify the resources for general use. A major rewrite was required; the new audience was very broad and the lines of work very diverse. The instructional design content had to be deleted. New and different examples of applying the thinking strategies were required for the books. To help the reader, each book required new learning activities. Exemplary workers in industry needed to field test and validate the content. Staff also needed to make major contributions to ensure the quality of each book. It took over six thousand hours to develop *The Exemplary Worker* series. In addition, industry has volunteered more than a thousand hours to field test and validate the content.

The Exemplary Worker series has many suggestions to help you not only be aware of your own thinking strategies but also help you to refine your strategies to achieve exemplary performance. You will also be better at mentoring others to perform better.

Gordon D. Shand
Edmonton, Alberta
Canada

Training Objectives

Upon completion of this book, you will be able to apply a thinking strategy to identify what is important to achieve a desired quality of results for materials and to minimize waste. You will be able to:

- Identify critical **input variables** of materials that define the quality of the materials used
- Identify the critical work and technical **process variables** that affect the quality of results
- Identify critical **output variables** that define the quality of results
- Identify the critical variables associated with:
  - refining and manufacturing
  - modifying materials
  - applying materials
  - packaging materials
  - storing and transporting materials
- Make changes to work and technical processes in response to:
  - desired changes in the quality of results
  - changes in input materials

Section 1
Introduction

This book is one of *The Exemplary Worker* series of books. Books in the series all focus on using critical thinking strategies to identify ***what is important*** to employees, the job, and the organization. Each book focuses on one of five domains (**PEMEO**):

P   People

E   Equipment

M   Materials

E   Environment

O   Organization

Within each book, loss and/or optimization (LO) are the main themes, hence the word LO-PEMEO™:

Themes	Books
L-P Loss to People (Safety)	*SafeThink* Use a structured thinking strategy to identify and predict hazardous situations.
O-P Optimize People's Performance	*WorkThink* Work effectively and efficiently.
LO-E Loss and Optimization of Equipment	*EquipThink* Use tools and equipment effectively and efficiently.
LO-M Loss and Optimization of Materials	*MatThink* Use materials effectively and efficiently.
LO-E Loss and Optimization of the Environment	*EnviroThink* Protect the environment.
LO-O Loss and Optimization of the Organization	*JobThink* Contribute to job and corporate performance.
LO-PEMEO Use thinking strategies for the workplace	*MetaThink* Integrate thinking strategies for exemplary performance.

The fundamental premise of LO-PEMEO is to *ask questions*. By asking yourself questions, you remain alert. By seeking answers, you continually learn and become more effective in the workplace and adaptable to changes. The big question is: *What questions should I ask?* The questions identified in LO-PEMEO help you to ask many of the right questions.

This book *MatThink* focuses on loss and optimization (LO) of materials. Low recovery rates when processing ore and damage to semi-finished materials are two examples of loss of materials. Using the right materials, maximizing the use of materials, and obtaining the desired results are three examples of optimizing materials. Generally, if you optimize materials, you reduce losses of materials and vice versa. The Job Aid for this book lists the key questions relating to materials.

The learning activities that are integrated into this book provide you with opportunities to apply critical thinking strategies to your job, workplace, and personal activities. Each learning activity relates to specific concepts addressed in this book. It is recommended that you complete each learning activity as you progress through the book.

The word *material* in this book refers to any raw, semi-finished, or finished material being mined, refined, modified, applied, packaged, stored, or transported. For many industries, such as agriculture, forestry, mining, manufacturing, and construction, materials are an important part of doing business. In these industries, optimizing the use of materials, minimizing material loss, and conserving energy are of fundamental importance in contributing to business success.

Two critical factors affect the optimum use and value of materials: **material quality** and **material waste**. Material quality and waste are closely related. For example, using low quality materials can result in additional waste. Controlling quality and minimizing waste can be difficult because there are many causes for adverse effects that degrade quality and increase waste. Most adverse effects on material are caused by PEMEO. For example:
- people (**P**) using poor material handling practices
- faulty equipment (**E**)

Materials can also create a Loss to PEMEO (i.e., L-PEMEO). For example:
- some materials break easily (e.g., glass), putting people (**P**) at risk of injury
- material waste can cause harm to the environment (**E**)

Where applicable, this book addresses how PEMEO impacts materials and how materials impact PEMEO (i.e., L-PEMEO).

Change Process

When working with materials, your main concern is to control changes through work and technical processes to achieve the desired result (quality). The following diagram shows the three stages of the change process:

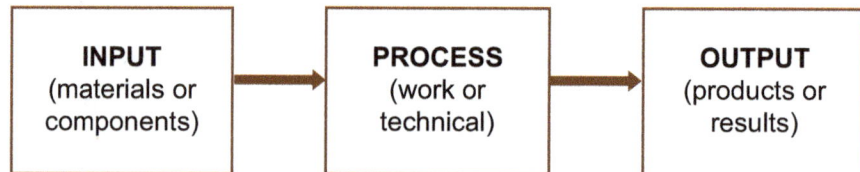

INPUT (materials or components)	→	PROCESS (work or technical)	→	OUTPUT (products or results)

In general, work and technical processes are used to change materials to produce a desired result (e.g., cut lumber to size, reshape metal, separate raw natural gas into various components, create a chemical change, assemble a cabinet). Generally, for a given project or customer, it is desirable for the quality of the output product to remain constant. Sometimes a change in the quality of the input materials creates the need to adjust the work or technical process to achieve the desired results.

In this book, the terms *work process* and *technical process* have different meanings:

Work process —People carry out a task to achieve a specific result. Often tools and equipment are used as part of carrying out the work. The task, the types of material, and the desired results determine the tools and equipment that will be used to carry out the work. For example: many types of lawn mower can be used to cut grass; many different tools and equipment can be used to cut metal; different types of earthmoving equipment can be used to move dirt.

Technical process—A stationary piece of equipment for manufacturing or processing performs a specific function to make a change to materials. The specific equipment dictates the tasks that people must do to operate the equipment. For example: start, monitor, adjust, shut down, and maintain the equipment.

1.1 Variables in the Change Process

Each stage of the change process involves variables:
- **input variables** define the quality of the materials used
- **process variables** create the change
- **output variables** define the quality of the product or results

Measuring Variables

Variables relate to properties, characteristics, or settings. Units of measure define the type of variable (e.g., temperature, pressure, volume, mass). Variables can be measured numerically or non-numerically:

- **numerical measurements** may be expressed as a unit value or ratio:
  1. units of measurement (e.g., temperature of 35°C, pressure of 100 kPa, dimension of 19 mm)
  2. ratio or percentage (e.g., 1:5 ratio of detergent to water, 7% asphalt by weight)
- **non-numerical measurements** are expressed in terms of our sense such as observations:
  1. a liquid appears cloudy instead of clear
  2. the surface of a board appears rough instead of smooth

Material properties: *scientifically defined qualities such as composition, density, and melting point*

Material characteristics: *size, shape, surface finish, type of break such as splintering that can often be controlled by the manufacturer*

Sometimes the measurement may not directly relate to the material itself, but to the method for handling the material. For example, the volume of material moved by an auger may be expressed in terms of auger speed (e.g., operate the auger at 225 rpm when transferring type 2 feedstock).

Input, process, and output variables have target values or optimal ranges. For example:
- the temperature of the adhesive must be above 15°C
- the pressure range of the gas entering the vessel can be between 500 kPa and 575 kPa
- the optimal flow rate is 30 litres per minute
- the optimal density of the output material is 120 g/cm^3
- the optimal composition of the product is 10% A, 30% B, and 60% C

NOTE

Material variables important to one job may be different from material variables important to other jobs. Sometimes, various trades have different concerns about the same material. For example, a pipefitter is concerned that the metal pipe does not leak and remains reliable under specific environmental conditions. An electrician is concerned that the same pipe conducts electricity and that, under specific conditions, an electrical current can be created in the pipe.

Classification of Variables

A variable can be classified according to its state and controllability:
- either static or dynamic (state)
- either controllable or non-controllable (controllability)

static variable	a variable that cannot be easily changed. For example, the composition of ore from a specific vein in a mine.
dynamic variable	a variable that can easily change. For example, the velocity of the wind, the number of patients coming to the hospital emergency.
controllable variable	a variable that you or technology can adjust. For example, the rpm of an engine, the size of plastic beads being manufactured.
non-controllable variable	a variable that cannot be easily adjusted. For example, the frequency of customer orders for a specific item.

NOTE

A variable may be static or dynamic depending on the job position or point of view. For example, at a specific maximum pressure, a piece of equipment will shut down. For the operator, that variable is static and does not change. An instrumentation technician, however, views the variable as dynamic because he or she can change the shutdown pressure and *fix* it at a new setting. For the operator, the new setting is *static* and he or she must work with the new setting until the technician makes a further change.

The two categories (state and controllability) of a variable combine to produce four possible *types of variable* as shown in the illustration on the following page.

Variable	Static	Dynamic
Controllable	static and controllable	dynamic and controllable
Non-Controllable	static and non-controllable	dynamic and non-controllable

Knowing whether a material variable can change and whether or not you can control the change is very helpful in doing your job well and responding effectively to changes. You must also understand the cause(s) or reason(s) for changes and how changes can *impact* PEMEO (i.e., LO-PEMEO).

The *type* of variable can have a significant impact on the loss and optimization of PEMEO, and on the decisions and responses that you make (within your limits of authority). The type of question you ask varies with the type of variable. For each of the following examples, the questions in italics are the **types of question** you might ask yourself to work effectively and efficiently and make *good* decisions for each *type of variable*.

Static and controllable variable—This type of variable remains the same throughout the work or process cycle but can be changed. Static and controllable variables are often associated with equipment. Usually production must stop to make the adjustments. For example, the pumping characteristics of a centrifugal pump are fixed, however, mechanics can modify the rotor (shave it) to change pumping characteristics.

The change to the pumping characteristics can affect the way you operate and the materials. Ask yourself questions about how the change to the pump affects the pump's operation and materials. Answers to the questions help you adjust the way you operate the pump to achieve production goals and minimize losses.

Ask yourself: *How does the modification of the pump affect pump performance? Is there a change to the minimum suction pressure? What is the new minimum suction pressure? Has the maximum operating speed (rpm) changed? What is the new maximum rpm? Has the maximum throughput changed? How much has the throughput changed? Has the discharge pressure changed and by how much? Are there new concerns about the change impacting PEMEO? Will the change to the pump affect how I plan transfer of the product? Do I need to allow for more time or less time than before to transfer a specific quantity of product? Can the changes to the pump affect product quality?*

NOTE

A variable may be static or dynamic depending on the job position or point of view. For example, at a specific maximum pressure, a piece of equipment will shut down. For the operator, that variable is static and does not change. An instrumentation technician, however, views the variable as dynamic because he or she can change the shutdown pressure and *fix* it at a new setting. For the operator, the new setting is *static* and he or she must work with the new setting until the technician makes a further change.

Dynamic and controllable variable—This type of variable can change throughout the work or process cycle, however, there is a means of controlling the variable to prevent damage to equipment and materials and to achieve the desired results.

For example, some products are made in batches. Materials or chemicals are mixed in specific ratios to achieve the desired results. The ratios and/or the composition of the input materials can be changed to achieve different desired results or products. When batches are changed, the equipment must often be purged of the previous materials and cleaned thoroughly. Any residue from the previous batch could contaminate the new batch, affecting the quality of the results. Asking yourself questions and seeking answers about the dynamic and controllable variables helps you work more safely and effectively.

Ask yourself: *What are the ratios of the materials for the specific batch? How much can each ratio change before the quality of results is affected? How do I know when the ratio of materials has changed? Can I control the ratios of the materials for the batch? What do I do if the ratios of input materials change? What can contaminate the input materials? What is the impact on the quality of results if there is contamination? How do I know the materials are contaminated? What do I do if the materials are contaminated? What is the impact on PEMEO if the input material ratios change or the materials are contaminated?*

Static and non-controllable variable—This type of variable does not change and cannot be changed. The composition of ore from a specific part of a mine remains relatively constant. Ore processing equipment may have to be adjusted to effectively process that specific ore composition. In building construction, workers must use the materials specified.

Ask yourself: *What property and characteristic variables of the material affect cutting and finishing? What variables must not change (e.g., finish)? What are the desired results? How can the material affect PEMEO? What equipment processing variables can be changed to obtain efficient and effective processing? What are the specific settings? Who makes the decision to change equipment operation? Who makes the changes? How are the changes made? After the changes are made, how long will it be before there is a noticeable difference in the specifications of the output products? What are the best methods for cutting the material? What methods are used to fasten the material? How can the material be damaged when handling and machining the material? How can the work affect PEMEO?*

Dynamic and non-controllable variable—This type of variable changes and cannot be controlled, for example, the variables associated with weather. Variables associated with weather include temperature, atmospheric pressure, humidity, precipitation, and wind velocity. In industry the composition, characteristics, and properties of raw materials may be dynamic and non-controllable. For example, molybdenum ore can vary in composition.

Ask yourself: *What is the impact of the variable on PEMEO? Can work, processes, and equipment be adjusted to safely and efficiently obtain the desired output results? What adjustments can be made? Who has authority to make decisions? What types of decision are made? Who carries out the adjustments? How are the adjustments made? If adjustments are made, what are the impacts on PEMEO? Is there a need for communication before, during, and after the adjustments are made? Is any documentation required? Who does the documenting and how is it done?*

NOTE

Often the type, size, and quality of large, stationary equipment for processing is static but some material, equipment, and processing variables may be controllable to achieve the desired output results.

The questions listed in italics are the types of question exemplary employees might ask when confronted with a specific work condition. Identifying variables, determining their types (change and control), and determining the impact that variables have on PEMEO are important thinking questions that you need to ask and answer to work effectively and efficiently. Within your roles and responsibilities and limits of authority, the answers to these types of question also help you make *good* decisions and respond effectively to both static and dynamic variables.

1.2 The Thinking Strategy

This book provides a **thinking strategy** to answer the question, *What is important about materials in my job?* The emphasis is not on the properties and characteristics of materials. Instead, this book focuses on the thinking strategy of *input, process, and output* variables so that you can work and respond to material changes more effectively. Using input, process, and output variables can help you:
- better understand how work and technical processes create change

- identify the material quality variables that relate to your job
- identify the critical input, process, and output variables that you need to pay attention to in your job
- determine which variables can and cannot change or be controlled
- define the quality of the input, process, and output variables
- adjust processes more effectively in response to input variable changes to maintain the desired quality of results
- adjust input variables and/or processes in response to a desired change in the quality of results
- deal with unfamiliar or different materials
- effectively communicate critical work and technical process issues to others

IMPORTANT

The *input, process*, and *output* thinking strategy helps you focus your thinking to identify variables that are important to your job. After you identify the important variables, you need to ask questions and seek answers about the variables so you can do your work safely, effectively, and efficiently.

This book applies the concept of *input, process*, and *output* variables to:
- refining and manufacturing materials
- modifying materials
- applying materials
- packaging materials

This book also addresses ways to minimize losses while storing and transporting materials.

MatThink™

Applying the Thinking Strategy to Optimize the Use of Materials

In your job, using the *input, process*, and *output* thinking strategy is very useful to determine the specific variables affecting material quality. When defining material quality, you must consider many variables such as material composition, consistency of mixture, strength, hardness, dimensions, shape, and finish. However, identifying *all* of the variables is neither practical nor useful. You only need to identify the quality variables that relate to your work or technical processes and affect your organization's goals.

The *input, process*, and *output* thinking strategy is used in three different sequences:

- *input–process–output*
- *output–process–input*
- *output–input–process*

Input–Process–Output

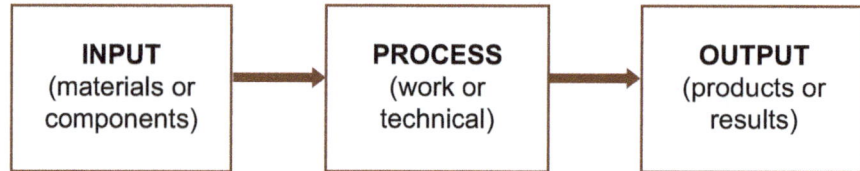

INPUT (materials or components)	→	PROCESS (work or technical)	→	OUTPUT (products or results)

In the illustration above, the key *input* variables that you must consider are those that:

- must be changed (e.g., separate undersized fruit from desired size of fruit)
- must **not** be changed (e.g., quality of desired size of fruit—no bruising)
- affect the efficiency of the process and equipment condition (e.g., equipment plugging up and wearing)

The key work or technical *process* variables are those that:

- create the change
- are changed
- affect equipment condition

The key *output* variables are material variables that define the quality of results.

NOTE

The output material of one process may become the input material for the next process of a system or assembly line.

Output–Process–Input

Sometimes, it is helpful to determine the critical material variables by working backwards. For example, examine the output quality variables first. Then examine the process to determine how the changes are made and the variables that are being affected. Finally, examine the input to identify the quality variables of the materials to determine which input variables must change and which input variables must **not** change.

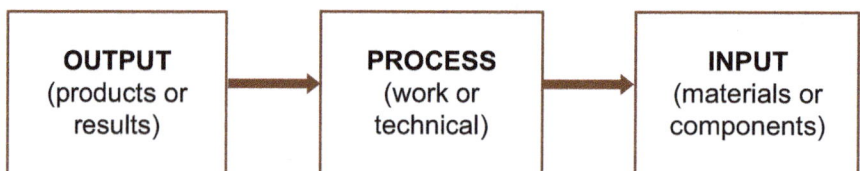

OUTPUT (products or results)	→	PROCESS (work or technical)	→	INPUT (materials or components)

Here is an example of using the *output–process–input* sequence:

output: A wall has a very bumpy and glossy finish and there appears to be a detailed pattern.

process: The effect was achieved by applying paint with a specially designed paint roller that creates the surface effect. The paint is very sticky, clinging to the paint roller and surface of the wall. As the roller moves, the paint is stretched away from the surface to form bumps. The speed of travel of the roller affects the height of the bumps. The person using the roller must maintain a constant speed to achieve consistency of bumps.

input: For the process to work well, the paint must be very thick and sticky and must dry very quickly.

Output–Input–Process

In some cases the output quality variables must be examined first and then the input quality variables examined to determine the best work process for getting the job done.

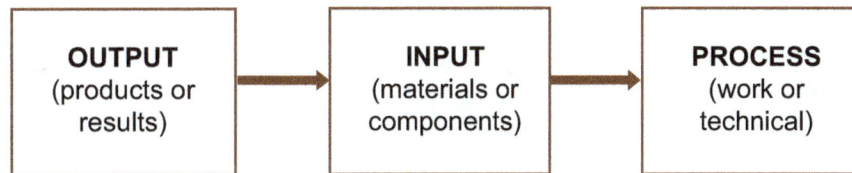

OUTPUT (products or results)		INPUT (materials or components)		PROCESS (work or technical)

This thinking sequence can also be helpful in understanding how a technical process works to achieve the desired results:
- Compare the variables between the output and input to determine which ones have changed and which ones have not changed.
- Investigate the process to determine how the change was achieved. Often the process creates the change through variables such as force, pressure, temperature, electricity, and chemical reactions. The process often involves two sets of variables: variables that *create change* and the material variables *that change*.

After you think through how the process works, it is often advantageous to apply different sequences of input, process, and output to refine your understanding of the variables important to your job. By doing so, you improve your ability to focus on critical variables so that you are more effective at optimizing the use of materials.

You can think of *input, process*, and *output* as a continuous loop.

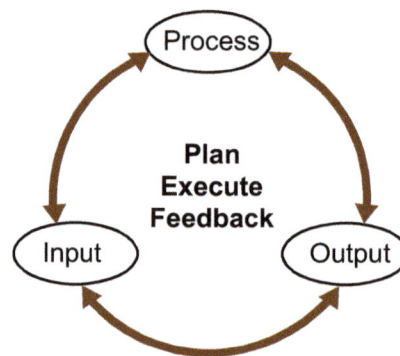

You can start at any point in the loop and go in either direction to plan, execute, and get feedback. The goal is to continually refine and improve safety, efficiency, and cost-effectiveness in achieving the desired results.

NOTE

The use of variables and the *input, process*, and *output* thinking strategy are very valuable to effectively and efficiently use materials and equipment. The major difference in applying the thinking strategy to materials rather than to equipment is the focus of the thinking. When applying the thinking strategy to materials, the primary focus is on the *material* input, process, and output variables. Only marginal attention is given to equipment variables. When applying the thinking strategy to equipment, the primary focus is on operating the equipment effectively to achieve the desired results without damaging the equipment or shortening its life. However, considerable attention must also be given to material variables because they affect equipment condition and the ability of the equipment to achieve the desired results. Refer to the book *EquipThink* for additional information.

Generic thinking process

This learning activity will help you refine your thinking skills to use the *input-process-output* thinking strategy to identify the technical and work process variables that affect materials.

1. Identify a work or technical process and answer the following questions. Examples of work processes are processing food using a food processor, cutting up sheets of plywood using a table saw, welding metal components to make a product. Technical processes involve using stationary equipment that refines materials or manufactures a product.

   The process is: _____

2. For your job, which *output variables* define the quality of results?

   _____

   _____

3. For your job, which are the *input variables* that affect the process and the quality of output results?

   _____

   _____

4. Which material variables are changed by the process?

   _____

   _____

5. Which material variables must **not** be changed by the process?

   _____

   _____

6. Which *process variables* change the materials?

   _____

   _____

7. What are the units of measurement for each material variable?

_____

_____

8. What are the units of measurement for the process variables that create the change?

_____

_____

9. What are the target values or optimal range for the material input and output variables?

Input material variables: _____

_____

Output material variables: _____

_____

10. Which input material variables are static? Dynamic?

_____

_____

11. Which *input* material variables are controllable? Non-controllable?

_____

_____

12. Which *process* variables are static? Dynamic?

_____

_____

13. Which *process* variables are controllable? Non-controllable?

_____

_____

Determining the Critical Variables

Critical variables are those variables that relate to your job and:
- define the quality of results
- define the quality of input materials
- create change
- are changed
- affect PEMEO (**P**eople, **E**quipment, **M**aterials, **E**nvironment, **O**rganization)
- are affected by PEMEO

To determine the critical quality variables for your job, you must:
- identify actual outputs, processes, and inputs
- identify the critical output quality variables
- identify the critical process quality variables. What conditions or actions are required to produce the product or result?
- identify the input material variable(s) that must be changed to achieve the desired output quality (product or result)
- identify the input material variables that must **not** be changed to achieve the desired output quality

To make the thinking strategy easier to use, work and technical processes involving materials can be grouped into four categories as shown in the following table.

Categories of Work and Technical Processes Using Materials	
Category	Process
Refining and manufacturing	Separating, extracting, combining, changing properties and characteristics
Modifying	Changing characteristics
Applying	Installing, attaching, coating
Packaging	Containing, protecting, identifying

As a starting point for identifying critical quality variables, you can categorize the process and then determine the type of change that is taking place or is required. Sometimes a process can be placed in more than one category. In this case, select both categories to identify the types of change that are taking place or are required. Then you can identify the variables that are important for your job and organization.

Each of these four categories is described in the sections which follow.

Section 4

Refining and Manufacturing

Category
Refining and Manufacturing
Modifying
Applying
Packaging

This section describes thinking strategies for identifying variables associated with refining and manufacturing. Strategies for refining processes are described first; strategies for manufacturing processes are similar, even though the outputs are somewhat more varied.

4.1 Refining

The purpose of refining is to separate materials such as minerals, oil, or gas into semi-finished or finished products, usable by-products, and waste. Composition is usually the most critical variable defining the quality of the refining outputs.

Output–Input–Process

To use the thinking strategy for identifying critical variables associated with refining, follow these steps:
- identify the composition quantity and quality specifications for the *output* products
- identify the *input* material variables that must be changed to achieve the desired *output* products

- identify the *input* material variables that must **not** be changed to achieve the desired *output* products
- identify the *process variables* that must be controlled to convert the input materials into the refined *output* products. Also determine:
  - how and why the *process variables* must be adjusted in response to changes in the input materials
  - how and why the process variables must be adjusted in response to desired changes in *quality* or *quantity* of the output products

Output–Process–Input

Alternatively, you can apply the *output–process–input* sequence:
- first identify the *output* variables
- then, identify the *process* variables that create the change
- next, identify those *input* variables that must be changed and those that must **not** be changed

Pages 23 and 24 provide examples of the *output–process–input* thinking strategy.

IMPORTANT

In refining processes, the following material usage objectives are important for the organization and the job:
- protect people and the environment
- obtain the desired quality of results
- maximize the use of materials
- minimize waste
- minimize energy consumption
- maximize efficiency of work
- minimize equipment wear and damage
- maintain competitive selling price

4.2 Manufacturing

The purpose of manufacturing is to convert raw and/or refined materials into marketable products such as steel, plastic, beer, or pharmaceuticals.

The conversion process often results in a change in the properties (such as tensile strength, malleability, chemical composition) and characteristics (such as surface finish) of the input materials.

The output products can be in several different forms:
- bulk materials that are used to make finished products (e.g., kraft for paper products, chemical feedstock, plastic pellets)
- semi-finished materials (e.g., rods, pipes, panels, bricks)
- finished materials (e.g., medicines, beverages, plastic components)

Example 1

Determine *output* variables

A blend of hydrocarbons is to be separated into two streams (output products): a blend of light, gaseous hydrocarbons and a blend of heavier, liquefied hydrocarbons.

Determine *critical* process variables

The critical process variable is pressure. A rapid drop in pressure causes some of the light, liquid hydrocarbons to rapidly change to a gaseous form. The heavier hydrocarbons remain as liquids. Controlling the process pressure controls the composition (quality variable) of each product stream (outputs).

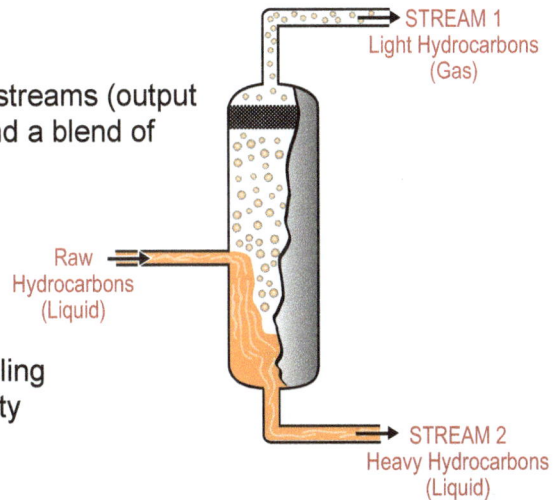

STREAM 1
Light Hydrocarbons
(Gas)

Raw
Hydrocarbons
(Liquid)

STREAM 2
Heavy Hydrocarbons
(Liquid)

Determine *input* variables that must be changed

Determine *input* variables that must not be changed

At the input of the process, the critical material variables are material composition, pressure, and temperature. The pressure and temperature of the materials must be such that the input stream is mostly liquid.

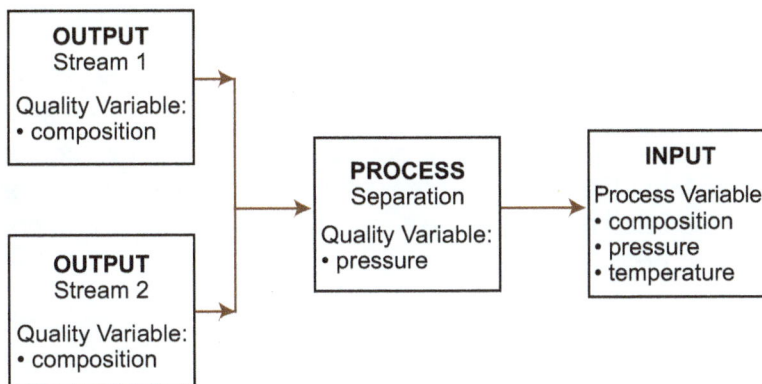

OUTPUT
Stream 1

Quality Variable:
• composition

OUTPUT
Stream 2

Quality Variable:
• composition

PROCESS
Separation

Quality Variable:
• pressure

INPUT

Process Variable:
• composition
• pressure
• temperature

In this example, the quantity (value) of each output stream depends on the percentage of each type of hydrocarbon in the input stream. That is, the value of the material outputs depends on the quality of the material inputs from the upstream process(es). You may or may not have control of the composition of the input material. However, you can control the process pressure to affect the composition (quality) of the output streams.

Example 2

Determine *output* variables

An oil emulsion (consisting of oil and water physically bonded together) is to be separated into two products: waste water and oil. The value of the output product depends on the ratio of water to oil in the emulsion. The larger the volume of oil and the smaller the volume of water in the emulsion, the greater the value of the output product.

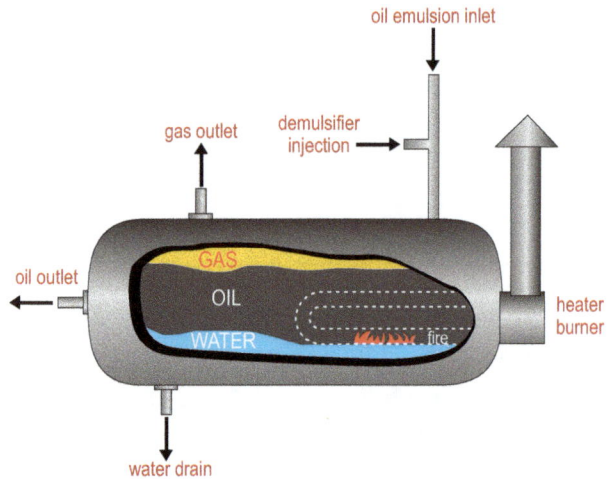

Determine *critical* process variables

The critical process variables are heat and a chemical demulsifier. Heat and the demulsifier break the physical bonds between the oil and water. Water, being heavier than oil, settles to the bottom of the vessel; the oil rises to the top. The quantity of heat and demulsifier determines the composition (quality variable) of the final product. Poorly separated oil still contains excessive water and gas.

Determine *input* variables

At the input of the process, the critical material variables are composition, tightness of the emulsion (degree of difficulty of separating the emulsion), temperature, and flow rate (volume).

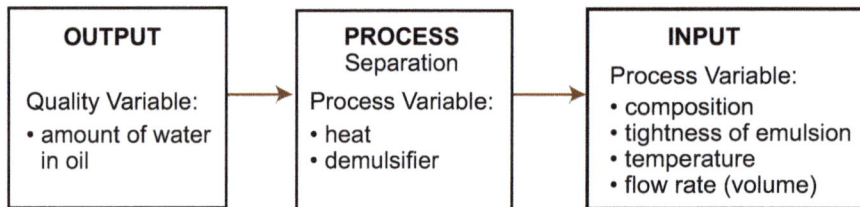

OUTPUT	PROCESS Separation	INPUT
Quality Variable:	Process Variable:	Process Variable:
• amount of water in oil	• heat • demulsifier	• composition • tightness of emulsion • temperature • flow rate (volume)

In this example, the amount of heat and demulsifier required to separate a specific volume of emulsion is dependent on the product temperature and tightness of the emulsion. Inadequate heat and demulsifier for processing results in poor separation. Excessive use of heat and chemicals is wasteful, creating unnecessary processing costs. By conducting tests, a company determines the optimum ratio and quantity of heat to demulsifiers required to cost-effectively achieve separation.

The quality of the incoming emulsion (input) determines the amount of heat and demulsifier used to obtain the desired outputs. You usually cannot control the quality of the incoming emulsion. If there is a change in input variables, separation may not be satisfactory. If the output product does not meet specifications, then you must know what actions are required to deal with the separation problem (e.g., increase heat, increase demulsifiers, slow down the flow rate, shut down the process). You must also know what to do with the rejected product.

The quality criteria of the outputs are specific to the product and its application. For example, quality in the food and beverage industry may include the following criteria:
- composition specifications including purity (absence of foreign material and biohazardous substances)
- volume
- temperature
- aroma
- texture
- flavor
- color
- attractive packaging
- timeliness

The thinking strategy sequence to identify critical variables is the same as for refining, either:
- output–input–process, or
- output–process–input

IMPORTANT

In manufacturing, the same material usage objectives apply as in refining:
- protect people and the environment
- obtain the desired quality of results
- maximize the use of materials
- minimize waste
- minimize energy consumption
- maximize efficiency of work
- minimize equipment wear and damage
- maintain competitive selling price

LEARNING ACTIVITY 2

Identify critical variables affecting material: refining and manufacturing

This learning activity will help you refine your thinking skills to identify, for your job, the critical variables affecting materials used in refining and manufacturing.

1. Identify a refining process and answer the following questions.

   The refining process is: _____

2. Fill out the following table identifying the consequences if the composition of the *raw* material changes significantly. Depending on the change in composition, some entries in the table may not apply.

Material Usage Objective	Consequences
People	
Environment	
Quality of results	
Maximum use of materials	
Minimum waste	
Minimum energy consumption	
Maximum efficiency	
Minimum equipment wear and damage	
Maintaining competitive selling price	

3. Identify a manufacturing process and answer the following questions.

   The manufacturing process is: _____

4. Fill out the following table identifying the effect on the process and quality of results if PEMEO functions poorly, behaves abnormally, or fails. Some entries in the table may not apply.

Domain	Condition, Action, or Event	Effect on Process and Quality of Results
People		
Equipment		
Materials		
Environment		
Organization*		

* Below is a list of some conditions, actions, and events that apply to the organization domain. Check off the examples that most likely apply to your job and work group.

- ☐ change in the work process
- ☐ shortage of staff
- ☐ lack of competent workers assigned to a specific task
- ☐ work group introduces new hazard
- ☐ third party fails to maintain equipment/facilities
- ☐ supplier changes standards or composition of components or materials
- ☐ unpredicted large customer order
- ☐ cancellation or delay of large customer order (e.g., may have excessive inventory on site)
- ☐ failure to communicate priorities
- ☐ lack of documentation
- ☐ administrative process inadequate for maintaining inventory
- ☐ failure to carry out routine safety or quality inspections
- ☐ failure to follow up on identified safety or quality deficiencies

MatThink™

Section 5

Modifying Materials

Category
Refining and Manufacturing
Modifying
Applying
Packaging

The purpose of modifying materials is to *refine* or *reshape* the materials for specific purposes. Industries, such as mining and construction, modify **raw materials** such as ore or gravel.

Modifying raw materials:
- sometimes produces the end result (e.g., grading and oiling dirt roads)
- may involve removing materials to prepare for another process (e.g., excavating and removing earth in preparation for construction activities)
- may involve removing materials for further processing (e.g., mining ore)
- may involve sorting, milling, crushing, screening, or washing materials

Industries such as construction and manufacturing modify (reshape and resize) **semi-finished materials** such as lumber, gypsum board (drywall), and textiles. The optimal use of semi-finished materials is affected by the quality of the materials, pattern design and layout, and selection and use of equipment.

IMPORTANT

When working with raw or semi-finished materials, the objectives that are important for the organization and the job are the same as for refining/manufacturing:
- protect people and the environment
- obtain the desired quality of results
- maximize the use of materials
- minimize waste
- minimize energy consumption
- maximize efficiency of work
- minimize equipment wear and damage
- maintain competitive selling price

The strategies for identifying critical variables associated with modifying materials are similar to those for refining in that they involve an analysis of outputs, processes, and inputs. Four stages must be considered when modifying raw and semi-finished materials to achieve the desired results:

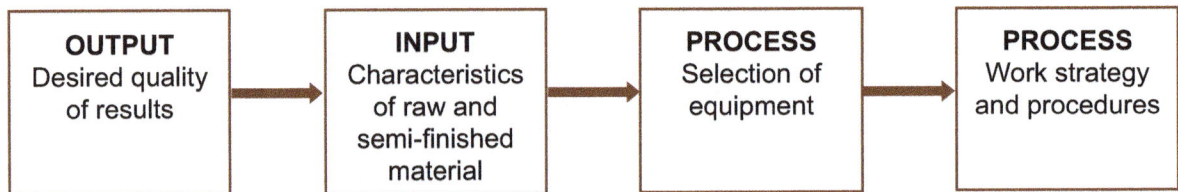

| OUTPUT
Desired quality of results | → | INPUT
Characteristics of raw and semi-finished material | → | PROCESS
Selection of equipment | → | PROCESS
Work strategy and procedures |

This section first describes modifying raw materials, then modifying semi-finished materials.

5.1 Modifying Raw Materials

Raw materials can be modified in two ways:
- by removing or reshaping the materials, as in excavating and grading
- by refining the materials, as in milling, crushing, screening or washing gravel or extracting diamonds from kimberlite (volcanic rock containing diamonds)

Removing and Reshaping Raw Materials

To get the desired results and efficiency of work when removing and reshaping raw materials, part of planning the work should include:
- defining the *output*: specifications that define the quality of results

- determining the *input*:
  - characteristics (e.g., hardness, looseness, coarseness) of the raw materials)
  - location and accessibility to the site of the raw materials
- determining the *processes*:
  - select the type and size of equipment that can efficiently and effectively handle the raw materials and access the site
  - determine the best approach (work strategy) to complete the job safely, effectively, and efficiently

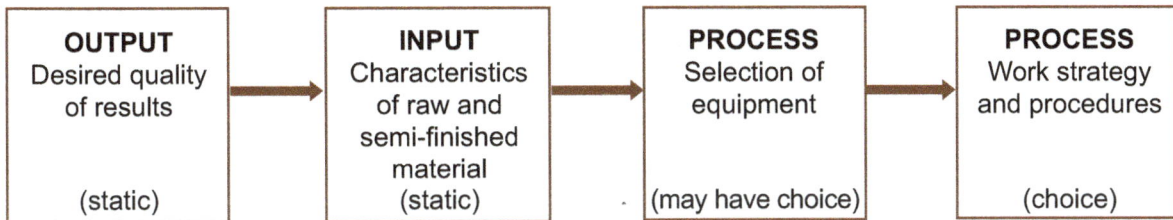

OUTPUT Desired quality of results (static)	→	INPUT Characteristics of raw and semi-finished material (static)	→	PROCESS Selection of equipment (may have choice)	→	PROCESS Work strategy and procedures (choice)

The decision-making process for excavating and grading is limited because the variables of the major factors are static (i.e., the desired result and the characteristics of raw materials). The variables that you may have some control over to achieve the desired results are selecting the:

- size and type of equipment
- work strategy and procedures

The selections of equipment and work strategies that you make depend on the:

- desired quality of results
- quality of input materials
- the specific project or amount of work to be done
- work environment

Output Variables—Output variables for removal of raw materials (i.e., excavating) are location, dimensions, shape, and finish. Shoring requirements and backsloping may have to be considered.

NOTE

The quality criteria for the removal of raw materials are the same for removing semi-finished or finished materials, for example, cutting openings in walls and floors. Refer to Section 5.2—*Modifying Semi-Finished Materials* for additional examples of quality criteria.

Input Variables—Input material characteristics such as hardness, looseness, and coarseness also influence the quality of the results, the selection of equipment, and the work strategies. For example, it is difficult to achieve:

- a vertical cut when digging a trench in loose sand because the sand caves in (dimension and shape are output variables)
- the desired shape and dimensions of the excavation when removing earth and minerals in frozen ground and permafrost
- a smooth grade when grading a road with gravel containing large rocks (finish is an output variable)

Process Variables (selection of equipment)—Using the right size of equipment for the application contributes to doing the work efficiently and effectively.

- **Undersized equipment** increases the time to do the work and, in some cases, reduces the quality of the results. For example, an undersized loader used for excavating pit-run is too light to dig through the materials easily and provide control for obtaining a level grade (finish is a desired quality of the results).
- **Oversized equipment** is difficult to maneuver in tight quarters and can cause damage to adjacent structures. For example, an oversized crawler tractor used for backfilling house foundations has limited mobility to obtain a level grade. The excessive equipment weight forcing the dirt against the walls can also shift or crack the walls.

Process Variables (work strategies and procedures)—Work strategies, procedures, and the skill of the operator affect the quality of the work. Through good mentoring and experience, equipment operators and tradespeople learn a variety of strategies to carry out work. For example, the pattern used to excavate a trench or hole depends on the location of the excavation. If the excavation is in an open area with no close-by buildings or trees to restrict the equipment movement, the operator has many choices as to where to start digging. However, if the trench or hole is close to buildings and trees, the operator must plan a pattern for excavating so that the equipment has adequate room to work.

Using safe and efficient procedures helps your organization achieve its objectives. Formal training, observing others doing similar work, and consulting with co-workers all help you learn to work effectively and efficiently. Being committed to developing exemplary operating skills refines your abilities to work effectively and efficiently.

Refining Raw Materials

In many situations, raw materials need to be refined. For example:
- gravel is screened and washed to get stones of specific sizes
- gravel made up of large rocks is crushed to get a consistent size of stone
- kimberlite is crushed and ground into small particles to recover the diamonds
- discarded trees are chipped and the chips are used as mulch

The stages for modifying raw materials and refining raw materials are the same.

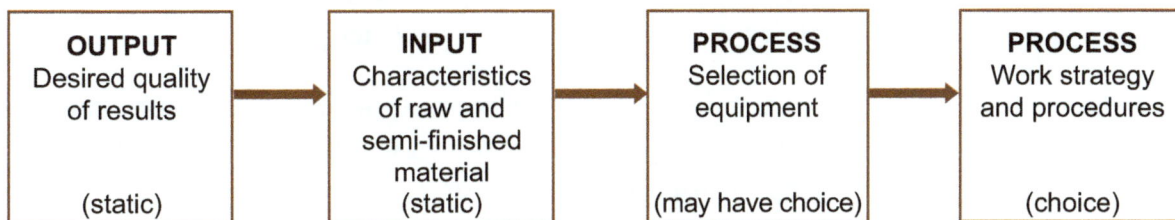

OUTPUT Desired quality of results (static)	INPUT Characteristics of raw and semi-finished material (static)	PROCESS Selection of equipment (may have choice)	PROCESS Work strategy and procedures (choice)

Output Variables—Usually, the output variables for refining raw materials are static. However, there are exceptions. For example, some types of wood chip are better for mulch because the wood decomposes faster than other types. Sometimes, chips that decompose slowly can be mixed with chips that decompose quickly (composition is a quality of results criterion) to obtain a satisfactory product. Different mixtures can be made after chipping (output). Alternatively, wood that decomposes slowly could be mixed with wood that decomposes quickly (input) before chipping to obtain a satisfactory product (output).

Input Variables—Input material characteristics such as composition and size affect the efficiency of the refining process and, in some cases, the quality of results. For example:

- The efficiency of screening and washing gravel to obtain specific sizes of stone can be affected by the size of the stones and the quantity and stickiness of foreign materials in the gravel.
- If most of the gravel being crushed into small stones consists of large rocks, the crushing process is difficult. The equipment undergoes excessive stress and additional energy is consumed.
- If kimberlite contains other types of rock, the grinding processes to recover the diamonds become inefficient.
- If all types of wood must be chipped, there is only limited control over the quality of results and the stress on equipment. Wood too large for the equipment may be cut into smaller pieces, used for another purpose, or discarded.

Process Variables (selection of equipment)—In large refining facilities, where equipment is stationary, there is limited choice in selecting equipment. If throughputs (output) vary, some equipment may be selected for use and other equipment may be shut down, depending on throughput and quality requirements, and product price.

For mobile or movable refining equipment such as wood chippers and gravel crushers there may be choices in equipment selection. Using the right size and type of equipment for the application contributes to doing the work efficiently and effectively. Throughput requirements and the quality of input materials are two key factors to consider when selecting equipment.

Process Variables (work strategies and procedures)—Often equipment can be adjusted in response to:

- changes in input material variables
- changes in the desired quality of results
- the stress on equipment

To operate equipment effectively, you need to know which equipment and material variables are dynamic and which variables are controllable. You also need to know:
- the reasons that equipment must be operated at a specific variable setting or within a specific range
- possible causes for variables to change
- indicators that variables have changed
- your response to changing variables

The book *EquipThink* provides in-depth information about operating equipment in ways to contribute to job and corporate performance.

Learn your company's procedures for operating the equipment and carrying out the tasks. For each specific process, learn the operating strategies and decision-making that contribute to your organization's goals. One way to learn to do the work and operate the technical processes is to assess the work planning process used by other workers. As part of the planning process, workers either assume or determine the variables that are controllable and those that are non-controllable for each stage:
- output: desired quality of results
- input: raw material properties and characteristics
- process: selection of equipment
- process: work strategies and procedures

If the product *quality (output) is static*, then workers need to control the dynamic variables for each of the other stages to obtain the desired product quality and minimize waste. If the *product quality can vary*, then workers can control the dynamic variables at each of the other stages to maximize throughput and minimize stress on the equipment.

NOTE

Minimizing risk of injury to people and harm to the environment applies to all work strategies.

5.2 Modifying Semi-Finished Materials

Semi-finished materials such as lumber, dry wall, panels, laminates, steel plates, pipe, ceramics, and textiles must often be changed in dimension, shape, and/or finish to serve a specific application. To achieve quality results, the same four stages used for modifying raw materials must be considered:

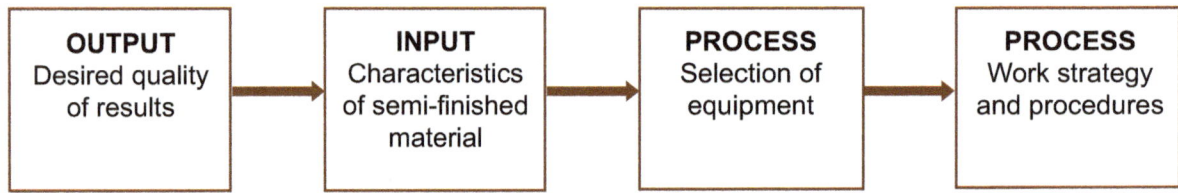

| OUTPUT Desired quality of results | → | INPUT Characteristics of semi-finished material | → | PROCESS Selection of equipment | → | PROCESS Work strategy and procedures |

Output Variables—Output variables for modifying semi-finished materials include properties, location, dimensions, shape, and finish to serve a specific application. For many types of material, the tolerances (accuracy) of the specifications may vary from application to application. In the modifying process, care must be taken to ensure that desirable properties and characteristics of the semi-finished materials are not changed. For example, strength and quality of finish of the semi-finished input materials must remain the same when the modifications are complete (output).

Input Variables—There are often many choices in selecting input materials, including:
- type of material (e.g., wood versus steel)
- properties (e.g., oak versus ash wood, tempered versus laminated glass)
- shape (e.g., channel versus box iron)
- dimensions (i.e., widths, lengths, and thicknesses to give the shape and size)

Sometimes input materials are of the required size and shape for immediate application. For example, wood studs are precut to length. Usually, however, semi-finished materials require modifications to achieve the desired results.

Material input variables can change, making carrying out the work and technical process more difficult and potentially changing the quality of results. For example:

- it is difficult to produce quality pieces for a cabinet when the wood warps excessively
- it is difficult to separate asphalt shingles in the bundle when the shingles have too much tar
- it is difficult to cut chipboard that has too little glue and too much foreign matter (e.g., metal fragments)

Process Variables (selection of equipment)—Often, there are choices in selecting tools and equipment to modify semi-finished materials. Selecting the right tool and/or equipment for specific semi-finished materials and for achieving a specific quality of result is beyond the scope of this book. Refer to the book *EquipThink* for detailed information about selecting tools and equipment.

You do not have a choice in selecting equipment in automated production facilities. However, you may be able to select specific components such as size and style of cutters on a milling machine to reshape the semi-finished materials efficiently and achieve the desired results.

Process Variables (work strategies and procedures)— Semi-finished materials often must be cut to size. Patterns and cuts must be laid out so there is minimum waste. Damaged sections of semi-finished materials must be avoided or located where the damage does not affect the quality of the product (e.g., will be covered by other materials). Sometimes damaged materials can be used for a different application where the particular type of damage is not an important quality of results variable.

NOTE

When materials are being cut manually, such as cutting plywood on a table saw, the patterns should be thought through and laid out before cutting to make work efficient. Efficiency can be improved by minimizing the number of times that:

- the equipment must be reset
- cuts stop in the middle of the material (i.e., when practical, the cut should run the entire length or width of the material)

Often materials must be machined to produce a component or part in large quantities. If automated production equipment is used, you have limited control over the variables that impact on the quality of results.

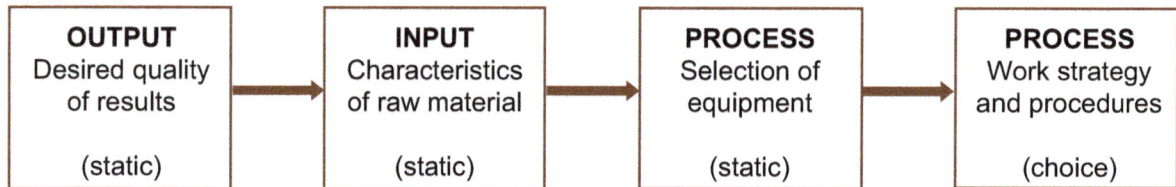

OUTPUT	INPUT	PROCESS	PROCESS
Desired quality of results	Characteristics of raw material	Selection of equipment	Work strategy and procedures
(static)	(static)	(static)	(choice)

As shown in this diagram, the output, input, and equipment (process) are static; you only have control over the work strategy and procedures. Some of the controllable variables might include the size and type of cutters used, depth of cut, speed of cutters, rate of travel, and the order in which the cuts are made. You would make these decisions on the basis of the type of material to be machined (static), the desired quality of results (static), and the capability of the equipment (static).

There can be several different causes for substandard quality. Low quality materials (e.g., lumber for cabinets warps, melting point of plastics may be too low) affect the strength, shape, fitness, and finish of the parts that are being made.

During the modification of semi-finished materials, the desired properties and characteristics can be changed:

Changing the desired properties

Hardness and strength may be a requirement of the output product. Processes such as grinding and cutting could produce enough heat to change the material's properties or, in some cases, melt the materials. Preventive measures, such as using oil to cool the material, reduce the risk of changing the material's properties. Specific equipment operating procedures may also reduce the heat being generated during the modification process.

Changing the desired characteristics

Semi-finished materials have characteristics that must not be changed during handling, storing, and modifying the materials. For example, Plexiglas™ plate has polished surfaces, a quality requirement of the output product. Any scratches or defects to the surfaces downgrade the product. Preventive measures, such as keeping working surfaces clear of debris and covering the Plexiglas™ surfaces with protective materials such as tape and paper, reduce the risk of damaging the surfaces.

Some semi-finished materials have properties that make it difficult to achieve the desired results without causing damage to the materials. The damaged products have to be rejected.

Meeting output specifications

Sometimes material properties, such as brittleness, increase the risk that the materials will chip or break during the modification process. Special tools, equipment, and operating procedures reduce the frequency of failure. Strategies such as scoring, notching, and starting cuts from both ends can be used to reduce the risk of damage during modification.

LEARNING ACTIVITY 3

Identify critical variables affecting material: modifying

This learning activity will help you refine your thinking skills to identify, for your job, the critical variables affecting materials that are being modified.

1. Identify a work or technical process that refines raw materials or modifies semi-finished materials and then answer the following questions. If you do not have such processes in the workplace, consider home projects. For example, you are cutting material to build a cabinet or

deck; you are replacing the floor with new materials; you have a hobby that requires materials to be cut and shaped.

The modifying process is: _____

2. For your job, which key *output* variables define quality of results?

   _____

   _____

3. What is the target or optimal range for each *output* variable that defines quality of results?

   _____

   _____

4. Which key *input* variables must be changed?

   _____

   _____

5. What is the target or optimal range of key *input* variables?

   _____

   _____

6. Which key *input* variables must **not** be changed?

   _____

   _____

7. What tools and equipment will be used to change the materials?

   _____

   _____

8.   What work strategy or technical process is used to make the changes?

_____

_____

9.   Which key *process* variables change the materials?

_____

_____

10. Which work or technical variables can downgrade the quality of results?

_____

_____

MatThink™

Section 6

Applying Materials

Category
Refining and Manufacturing
Modifying
Applying
Packaging

Applying materials includes installing, attaching, or coating:

- semi-finished materials such as lumber, panels, siding, steel plates, textiles, paint, and wallpaper
- components such as dental braces, zippers, and electrical and plumbing hardware

Four stages are considered when installing or applying **semi-finished materials** and **components** as shown in the drawing below:

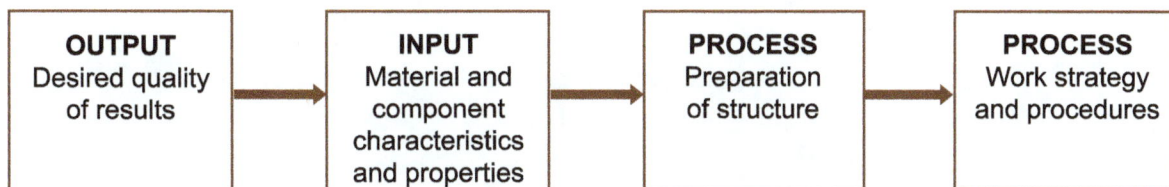

OUTPUT Desired quality of results	→	**INPUT** Material and component characteristics and properties	→	**PROCESS** Preparation of structure	→	**PROCESS** Work strategy and procedures

For many work assignments, specific types of semi-finished material and components are selected in advance to achieve the desired results. To control the quality of results, work efficiently, and minimize rework time, you must first eliminate defective materials. To maximize the use of the materials, you must determine the properties and characteristics of the materials that can create difficulties in achieving the desired results. Using the right materials or components and applying them correctly and in the correct location minimizes material waste and rework time.

Output Variables—Material quality *output* variables include location, fit, bond, material characteristics, and material properties.

For components used by tradespeople, the quality of components and their installation may be controlled by regulations and codes. Regulations and codes address issues such as component function, component application, durability, safety, and environment.

Customer expectations may be higher than the standards set out by regulations and codes. A customer's criteria for quality may be driven by his or her organization's corporate goals.

Materials and components must be positioned, fitted, or mounted according to quality criteria, which could include:
- being level
- being square
- correct alignment in relation to other objects (e.g., patterns in flooring and wallpaper, buttons to button holes, washroom mirrors to sinks, engines to pumps)
- being fitted against adjacent material within specific tolerances (e.g., size of gap between panels, flushness of electrical switch plates to walls)
- smoothness of exposed corners and edges
- smoothness and evenness of surface finish
- strength of mountings
- allowance for expansion, contraction, shifting, and settling of materials

In some cases, the general location of components and semi-finished materials such as fixtures, panels, and siding is pre-determined and shown on blueprints. In other cases, the specific location of components is left to the installer's discretion. The location of fixtures and components is determined by several factors including the need to:
- maximize the function of the fixture or component

(e.g., locating light fixtures to illuminate the desired area and minimize shadows)
- minimize the potential for damage to materials, fixtures, and components (e.g., gas sensors that must be close to the floor should be mounted away from high traffic areas; walls finished early in a project may be scarred before the project is completed)
- ensure safety (e.g., locating electrical outlets close to the load to prevent running extension cords across traffic areas)
- ensure accessibility to critical parts of the components or fixtures (e.g., access to the pilot light and shut-off valves of a hot water heater)
- minimize interference with other components and structures (e.g., a shelf mounted behind a door prevents the door from opening fully)
- maximize the use of space (e.g., locating utilities such as water heaters and furnaces close to walls and adjacent to each other)
- consider aesthetics

Often the quality of results is affected by the quality of the frame or structure to which the components or semi-finished materials must be attached. For example, wall studs must be square and straight so that drywall, panels, or cabinets will fit flush and true. In some cases a crooked structure will cause brittle finishing materials to break and components such as cabinets to distort and become misaligned during installation.

Input Variables—Semi-finished materials often must be modified before they can be applied (see Section 5.2).

Properties and characteristics of a given type of material or component can vary from manufacturer to manufacturer. Suggestions from other workers who use the materials and components can be valuable in selecting materials that contribute to:
- achieving the desired quality of results
- minimizing waste
- preventing harm to people
- protecting the environment
- improving work efficiencies

Regulations and codes often dictate the properties and characteristics of components (e.g., electrical components). When renovating, refurbishing, or repairing, it is often desirable to replace components in kind. Sometimes it may be wise to check with the manufacturer or supplier to determine if the materials or components are appropriate for the application.

Process Variables (preparation of the structure)—
Components such as shelves and hand rails must be mounted to a strong structure (in this application, one criterion of quality is function: to provide support). Additional supports may have to be added to the structure to provide a strong anchor for the component. For example, additional support would be required if a hand rail anchor does not align with a stud.

In cases where glues, adhesives, and cements are used to attach materials, the main structure must be prepared. Surfaces must be clean and smooth.

Process Variables (work strategy and procedures)—
Careful planning contributes to minimizing waste, the efficiency of doing the work, and the effectiveness of results.

The properties of the materials must be known so that work methods can be used to minimize damage to the materials. For example, handling and installation activities could break brittle materials or scar or dent soft materials. Generally, materials that are difficult to handle (e.g., heavy, awkward, very small, fragile) have a greater potential for damage and waste. To minimize waste, damaged materials can sometimes be used in locations where the damage does not affect the quality of results.

Some materials, such as tongue and groove lumber, flooring, wallpaper, ceramic tile, and fabrics have specific shapes or patterns. When these materials are cut to fit the application, the removed parts generally can only be used to start or finish a new section. Careful planning of material layout can sometimes reduce waste by ensuring that the cut off parts are of adequate size, shape, or pattern to be used at a different location.

Regulations and codes may dictate how some materials and components are to be installed. In addition to meeting regulations and codes, there may be opportunities to maximize the efficiencies of installation and meet quality issues important to customers. Refer to the book *WorkThink* for more information.

Attachment Methods

Many different methods, including mechanical fasteners, adhesives, and welds, are used to attach materials and components:

- Mechanical fasteners come in many shapes and forms such as nails, gang nails, screws, bolts, rivets, joist hangers, plastic ties, metal clips, and hinges.
- Many different types of glue, adhesive, and cement are used to attach materials and components which have diverse characteristics and properties. Metals and plastics often have properties that make them suitable for welding or fusing.

For attachment methods, the criteria for quality of results include:

- strength of connection
- durability in the environment
- minimized damage to the attaching structure and materials during installation
- aesthetics

For attachments, the thinking process to achieve the desired quality of results involves:

- specifying the quality criteria of results (*output*)
- determining the best method for making the attachments (if there are choices) (*process*)
- selecting the correct quality of attachments (e.g., bolts come in different grades of strength, cements can have different properties and characteristics, welding rods have different properties) (*process*)
- determining the critical process variables that affect the results (*process*)
- identifying the critical input variables affecting the quality of results (e.g., surfaces must be clean and smooth) (*input*)

Mechanical Fasteners

Mechanical fasteners that have threads must be tightened to a specific tension. The amount of tension is determined by the torque used to tighten the fasteners. Excessive torque on the fasteners can result in premature fatigue failure; the materials being attached can experience uneven stress and damage due to excess compression. Under-torqued fasteners can result in loose and dysfunctional contact.

Glues, Adhesives, Cements

In cases where glues, adhesives, and cements are used to attach materials, the main structure must be prepared. Surfaces must be clean and free of debris and foreign materials. Surface imperfections such as bumps and cavities must be corrected. Some adhesives require that the attaching surfaces be roughened or treated with chemicals to improve adhesion. When the composition of the two attaching materials is very different, two different adhesives may have to be applied. The first adhesive is applied to one material to provide a compatible surface to which the other adhesive can bond (e.g., attaching braces to teeth). Some adhesives consist of two components that must be mixed before applying. A chemical reaction takes place to create the bonding agent.

Bonding materials such as adhesives and cements have an open time. That is, once the bonding material has been applied to a surface, there is a time limit (a few minutes to several hours) in which the second component must come in contact with the first component. Components not positioned within the time limit results in poor bonding and fit. After contact, heat, ultra-violet light, and/or pressure may have to be applied for a specified time for the bonding material to adhere effectively. In the case of contact cement, the cement is applied to both contact surfaces and allowed to dry. After drying, there is a time limit in which the components must be positioned. After bonding, some adhesives and cements require a curing period to develop full strength.

Large projects using adhesives and cements need careful planning. For efficiency, the bonding material is applied to a large area. The size of the area is selected such that the last component will be positioned just before the bonding material reaches its open time limit.

WARNING

When working with bonding agents, always check the Material Safety Data Sheet (MSDS) of each product to identify the hazards and precautions. Non-controlled products do not have MSDSs and only give limited warnings. Excessive exposure can potentially cause allergies and other reactions.

Primers, Sealants, Paints, and Finishes

For primers, sealants, paints, and finishes, surface preparations are similar to those for glues, adhesives, and cements. If more than one coat is to be applied, the first coat must be dry and possibly roughened. Some products, such as urethanes, do not adhere well to themselves once they have dried thoroughly. Second and consecutive coats must be applied within a specific time to improve adhesion.

Sometimes the applied materials need to be thinned with a solvent so that the materials spread evenly and adhere properly to the surface. The correct solvent must be selected and only the minimum quantity added to the compound to achieve the desired results. Excess quantities of solvent can change the characteristics and properties of the compound, potentially reducing the quality of the results. The quantity of compound applied to the surface can also affect the quality of results. Excess or insufficient coverage of finishing compounds affect bonding ability and surface finish.

WARNING

When working with surface finishing products, always check the MSDS of each product to identify the hazards and precautions.

Control of Processes and Quality

If you work in an industry such as construction, renovation, or repair, you may have significant control over the work process of fitting and installing semi-finished materials and components. You also have some control over the quality of results. However, if you work on an assembly line, you have limited control over the work processes and the variables that contribute to the quality of results. In the figure below, quality of results, selection of equipment, and characteristics of materials are static and cannot be changed.

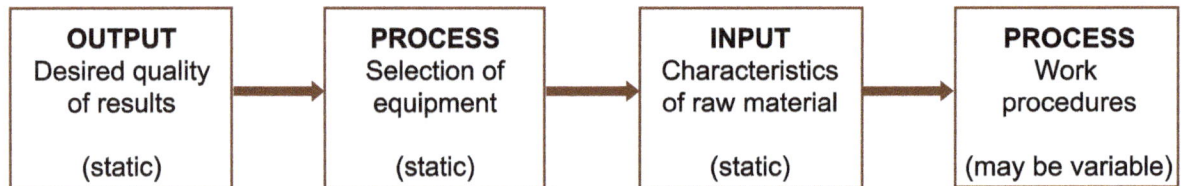

OUTPUT Desired quality of results (static)	→	**PROCESS** Selection of equipment (static)	→	**INPUT** Characteristics of raw material (static)	→	**PROCESS** Work procedures (may be variable)

Often the steps of the procedure are static, but the quality of performance of each step can vary. The use of aids such as jigs and hold downs help improve efficiency and control quality. Poor ergonomics and fatigue can contribute to poor performance and a decrease in product quality. Quality criteria that can be affected by poor performance include fit, function, and aesthetics.

IMPORTANT

When applying materials, the following objectives are important for the organization and the job:
- protect people and the environment
- obtain the desired quality of results
- maximize the use of materials
- minimize waste
- minimize energy consumption
- maximize efficiency of work
- minimize equipment wear and damage
- maintain competitive selling price

Summary of Applying Materials

Think Ahead

For many situations in which materials are being used, it is important to think ahead and plan the work to achieve the desired results and efficiency. In some cases, work must be coordinated with other work groups to minimize damage and waste of materials.

Critical variables important to your job can be identified by assessing the work using one of the two following thinking strategies:

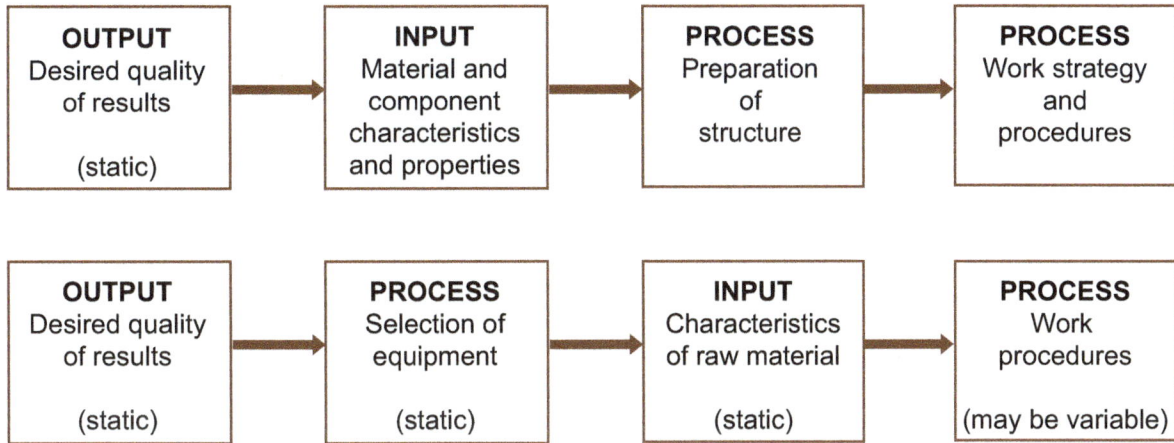

OUTPUT Desired quality of results (static)	INPUT Material and component characteristics and properties	PROCESS Preparation of structure	PROCESS Work strategy and procedures

OUTPUT Desired quality of results (static)	PROCESS Selection of equipment (static)	INPUT Characteristics of raw material (static)	PROCESS Work procedures (may be variable)

Employees doing specialized work, such as those in the trades, often use specific techniques and strategies that maximize the use of materials, minimize damage, and make work easier and more efficient. Specific work habits, such as housekeeping and storing materials in a safe location, contribute to the safety of other workers and minimize material damage.

LEARNING ACTIVITY 4

Identify critical variables affecting material: applying

This learning activity will help you refine your thinking skills to apply materials effectively.

1.   Identify a work or technical process where components or semi-finished materials are applied and then answer the following questions. If you do not have such processes in the workplace, consider home projects. For example, tiling a bathroom or floor, painting or wallpapering a room.

The application process is: _____

2. For your job, which key *output* variables define the quality of results?

_____

_____

3. What is the target or optimal range for each *output* variable that defines quality of results?

_____

_____

4. Which key *input* variables affect the quality of results?

_____

_____

5. Which key *input* variables of the material receiving the application must be satisfied to obtain quality of results?

_____

_____

6. What tools and equipment will be used to apply the materials?

_____

_____

7. What work strategy or technical process is used to apply the materials effectively and efficiently?

_____

_____

8. What work or technical variables can downgrade the quality of results?

_____

_____

Section 7

Packaging Materials

Category
Refining and Manufacturing
Modifying
Applying
Packaging

The purpose of packaging is to contain, protect, and identify products. Products can be gases, liquids, or solids, or a combination of two or three of these states.

There are four stages to consider when packaging materials:

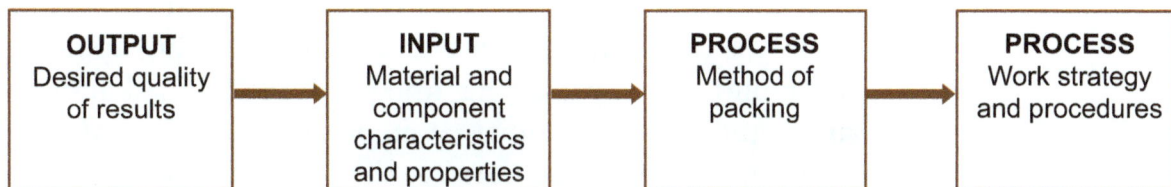

OUTPUT Desired quality of results	→	INPUT Material and component characteristics and properties	→	PROCESS Method of packing	→	PROCESS Work strategy and procedures

Output Variables—The type of product to be packaged determines the criteria for the quality of results:

- bulk (e.g., natural gas, crude oil, potash, wood shavings, wheat)
- single product (e.g., pulp bales, household appliance, door lock, electric fan, computer monitor)
- multiple types of product at one location prepared for shipping (e.g., auto parts, electrical components, pharmaceuticals, groceries)

53

Packaging must be sufficient to protect the contents from potential damage caused by PEMEO. Some contents are harmful. The packaging must withstand potential damage to prevent the product from breaking through the packaging and affecting PEMEO (i.e., L-PEMEO).

The criteria for quality of results are different for each type of product and packaging used. Variables include:

Variables for Selecting Methods of Packaging	
• physical properties of product • characteristics of product • control of product damage • control of environmental contamination • required quantities	• method of storing and transporting • safety • marketing, labeling, and coding requirements • application requirements • customer expectations

Input Variables—Input variables include the types of material being packaged: bulk, single product, multiple types of product. The quantity to be transported also affects the type of packaging. For example, propane (a liquid fuel under pressure) may be contained in small cylinders used for torches and barbecues or in large vessels mounted on trucks or railcars.

The properties and characteristics of materials (listed in the previous table) also affect the type of packaging appropriate for handling, transporting, and storing. The packaging, inks, and adhesives must be made of materials that do not contaminate or cause harm to the contents and will not deteriorate under transportation and storage conditions.

Often packaged materials are shipped and stored on pallets. The wrong packaging could fail to protect and contain its contents when handled, transported, or stored.

Process Variables—While you may have little control over the type of packaging, you may be able to adjust the packing method and work procedures to achieve the desired quality results. Inferior or substandard packing methods and procedures can have the potential to harm PEMEO.

LEARNING ACTIVITY 5

Identify critical variables affecting material: packaging

This learning activity will help you refine your thinking skills to identify, for your job, the critical variables affecting the packaging of materials.

1.   Give an example of a product that was damaged due to poor packaging and explain how the packaging could be improved.

     _____

     _____

2.   Give an example of a product that had quality packaging and explain why you think the packaging was excellent.

     _____

     _____

3.   Give an example of a product that had excessive packaging that resulted in waste and potential harm to the environment.

     _____

     _____

MatThink™

56

Storing and Transporting Materials

Both direct and indirect losses can occur while storing and transporting materials. Direct losses occur as a result of damage to materials. Indirect losses include reduced productivity because of material shortages and an increase in the cost of insurance. Extra costs also occur when purchasing and storing excess inventory.

8.1 Storing Materials

Material Damage

Material damage can be defined as any change to the material that degrades the material's quality. Generally, damage results in a change in the properties or characteristics of materials and components. Both the environment and work actions can lead to damage of stored materials:

- inadequate or excessive light
- inadequate or excessive temperature
- inadequate or excessive humidity
- inadequate or excessive ventilation

- exposure to:
  - dust
  - water
  - solvents
  - oils (including fingerprints)
  - static electricity
  - foreign objects (e.g., wood chips between sheets of plywood cause scarring)
- excessive stress from the weight of the material (e.g., material stacked too high; some materials sag if suspended horizontally at both ends for a long time)
- excessive impact force
- excessive vibration

You can take several measures to minimize damage to materials and components stored outdoors:
- use covers to protect the materials from the elements
- avoid locating the materials in low-lying areas where water can pool
- elevate the materials to minimize exposure to soil, moisture, rodents, and insects
- slightly slope flat materials to prevent water from pooling on the top surfaces
- position openings on containers so that pooled water cannot leak through the access openings and contaminate the product (e.g., slope drums slightly so that the two access openings—bungs—are away from the low side)

Locate material stored outdoors to permit easy access. In some cases, the stored material can be an eyesore. Screens and fences can prevent complaints from neighbors or other businesses.

Several different factors associated with inventory can cause direct or indirect losses (shrinkage):
- product shelf-life
- inadequate product quality
- excess product quality
- decentralized stock controls
- excess quantity (overstocking)
- shortage of critical items
- fluctuating seasonal demands

Some materials, such as carbon-zinc batteries and fresh foods, must be used before a specified time. Rotate stock so that older items are used first.

Quality of Stock

As inventories are depleted, new stock must be ordered. The quality of the replacement stock can create both direct and indirect losses. If the replacement materials are of lower quality, several types of loss can occur:

- Material is wasted because some of the material is not satisfactory for the application.
- Safety can be compromised because the replacement material behaves differently than the original when used.
- More time may be required to carry out work.
- Components may fail prematurely, creating downtime and additional maintenance.

In many cases, the direct cost of materials is proportional to quality. The cost of high quality materials may or may not be justified. Input from different departments such as accounting, purchasing, marketing, and the user group may be needed to select materials that give the organization the most value for the investment. Materials should be purchased from reputable suppliers to protect against purchasing substandard materials and counterfeit parts.

Correct Amount of Inventory

Keeping the *correct* amount of inventory is critical to minimizing losses. Excess inventory is costly to carry. Organizations with several locations can have excess inventory at some locations and shortages at other locations. A shortage can result in lost sales or production. A shortage of critical parts for equipment maintenance can result in weeks or months of lost production until the parts can be delivered.

A change in consumer preferences, regulations, and advances in technology can result in material waste. An organization can end up carrying excess inventory of outdated materials and components.

Seasonal demands can also have an impact on inventory requirements, making predicting demand even more difficult. Many organizations improve inventory control with computer software, universal product codes (UPC), and Radio Frequency Identification (RFID) to track consumption, orders, inventory, and costs. The software may be a standalone package used at one location or a multi-user package accessible from many different locations. The value of the software for controlling inventory depends on the quality of the input data and its interpretation.

Just-in-time delivery practices can be used to maintain a minimum amount of inventory to reduce the financial costs of carrying inventory. Arrangements are made with suppliers to deliver the products just before they are needed. Whenever possible, excess items are returned to minimize carrying costs.

Another trend for optimizing material use and minimizing waste is to apply conservation practices—reduce, reuse, and recycle.

8.2 Transporting Materials

During handling and transportation, materials can be damaged by extreme temperatures, vibration, rubbing, and impact forces. Often materials are put on pallets and shrink wrapped for easier handling and to prevent movement. To reduce rubbing and impact forces, dunnage (e.g., off-cut or spare pieces of scrap wood or cardboard) is often placed between materials.

Materials can be dangerous, potentially affecting people's safety, damaging equipment and other materials, and harming the environment. Governments have developed regulations and standards for hazardous materials. In Canada, when transporting dangerous goods, Transportation of Dangerous Goods (TDG) or equivalent legislation requirements must be met. When handling hazardous materials, the Workplace Hazardous Materials Information System (WHMIS) must be used. Regulations and standards may vary from one country to the next. Governments are negotiating to develop harmonized standards that apply globally. Hazardous materials are addressed in the book *SafeThink* which is a structured critical thinking strategy to identify and predict hazardous situations.

Identify critical variables affecting material: storing and transporting

This learning activity will help you refine your thinking skills to identify, for your job, the critical variables affecting the storage and transportation of materials.

To answer the following questions, use examples from work or from home.

1.   Give an example of a product that had to be discarded because its shelf-life had expired.

_____

_____

2.   Give an example of a product that degraded because it was not stored in a suitable environment (e.g., too hot).

_____

_____

3.   Give an example of a product that was damaged because it was stored poorly or located in a traffic area.

_____

_____

4.   Give an example of a product that was contaminated because it was not protected from the surrounding environment.

_____

_____

5. Give an example of an excessive quantity of product that could not be used within a reasonable amount of time.

_____

_____

6. Give an example of a shortage of product that created lost time or extra effort to replenish.

_____

_____

7. Give an example of materials being transported that spilled and could have put people at risk or contaminated the environment.

_____

_____

Section 9
Summary

The thinking strategy to use materials effectively and efficiently involves identifying and analyzing the critical variables associated with outputs, processes, and inputs.

When your job involves materials, look for opportunities to optimize the use of materials and minimize waste. If you are dealing with a technical or work process that uses or produces materials, use the *input–process–output* thinking strategy to identify the critical variables that relate to your job.

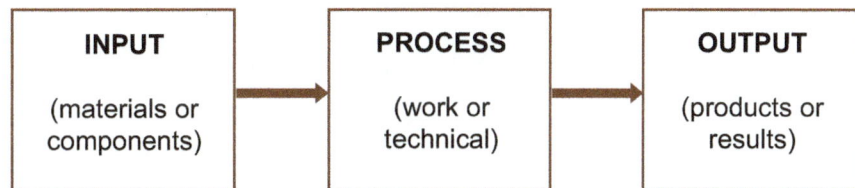

INPUT	→	**PROCESS**	→	**OUTPUT**
(materials or components)		(work or technical)		(products or results)

The following steps provide a summary of the thinking process.

1. Identify actual *outputs, processes,* and *inputs.*

    Often the process and the purpose of the process can be used to determine actual outputs, processes, and inputs. For example, the statement, *an amine system removes sour gas from a raw gas stream to produce sweet gas* indicates the following:
    - *output:*   sweet gas and sour gas

- *process**: remove sour gas from raw gas using amine
- *input*:   raw gas

The statement, *the metal lathe is used to machine the brass rods into ornamental bells* indicates the following:
- *output*:   ornamental bells
- *process**: machining using a metal lathe
- *input*:   brass rods

* For many material uses, there may be more than one work or technical process as in preparing structures and applying materials.

2.   Identify the critical quality *output* variables and their specifications. For example, percentage of water in natural gas, smoothness of finish of wood panels, tolerance of bearing dimensions, dimensions of an excavation.

To help identify the types of critical variables that may be important for defining the desired output, use the classification scheme below. This classification scheme separates technical and work processes into four categories.

Categories of Work and Technical Processes Using Materials	
Category	**Process**
Refining and manufacturing	Separating, combining, changing properties and characteristics
Modifying	Changing characteristics
Applying	Installing, attaching, coating
Packaging	Containing, protecting, identifying

Within each category there are common types of quality variables that affect the optimal use and value of materials. Refer to the specific sections on refining/manufacturing, modifying, applying, and packaging for suggestions of the types of quality variables that may apply to the particular process.

3. Identify the critical *few process* variables and their specifications, that is, the conditions or actions that change the input material to produce the output material.

4. Identify the *input* variables that are changed and their specifications.

5. Identify the *input* variables that must be controlled for the process to work and to achieve the desired results.

6. Identify the *input* variables that must **not** be changed.

7. Determine whether each critical input, process, and output variable is static or dynamic.

8. Determine whether or not you can control each critical variable.

The following table summarizes steps 1 through 8.

Action Step	Output	Process	Input
Identify outputs, processes, and inputs	• result • product	• technology • work	• raw materials • semi-finished materials • components
Identify critical variables	• quality • quantity • time • timeliness	• method of causing change • variables that change	• variables that must be changed • variables that must **not** be changed
Determine specifications of variables	• static/setpoint • optimal range	• target • optimal range	• target • optimal range
Determine if variables are static or dynamic	• static • dynamic	• static • dynamic	• static • dynamic
Determine if variables can be controlled	• controllable • non-controllable	• controllable • non-controllable	• controllable • non-controllable

9. When carrying out tasks, look for ways to effectively achieve results, control damage and waste, and improve efficiencies.

10. Preplan your responses to changes in variables. Always keep corporate goals and policies in mind so that your responses contribute to the organization and are within your authority to take action.

The following tables summarize the key issues to consider when selecting packaging methods and procedures.

Variables for Selecting Methods of Packaging	
• physical properties of product	• safety
• characteristics of product	• marketing, labeling, and coding requirements
• control of product damage	• application requirements
• control of environmental contamination	• customer expectations
• required quantities	
• method of storing and transporting	

Causes of Damage to Stored Materials	
• inadequate or excessive light	• exposure to rodents and insects
• inadequate or excessive temperature	• exposure to static electricity
• inadequate or excessive humidity	• exposure to foreign objects
• inadequate or excessive ventilation	• excessive stress from the weight of material
• exposure to dust	• excessive impact force
• exposure to water	• excessive vibration
• exposure to solvents	
• exposure to oils	

Factors Associated with Inventory that can Cause Losses	
• product shelf life	• excess quantity
• inadequate product quality	• shortage of critical items
• excess product quality	• fluctuating seasonal demands
• decentralized stock controls	

Job Aid

Using this Job Aid

Identifying the critical variables for the materials that you use or change in a process:
- helps you to focus your efforts to efficiently control the quality of results
- provides valuable information when mentoring others

Instructions

1. select a material. Determine your responsibility for working with that material:
   - producing a product by refining, manufacturing, or modifying material
   - using the material to make something

2. For each set of questions, check off the questions that are most important to achieve the desired results.

3. For the variables you have identified, make notes explaining the specific variables and their standards, setpoints or ranges.

4. Explain what can go wrong and how to respond.

MatThink™

Notes

Materials: _____

Purpose of materials: _____

Work or technical process: _____

_____

_____

_____

_____

_____

_____

_____

_____

_____

_____

_____

_____

_____

_____

_____

_____

_____

_____

_____

_____

_____

_____

_____

_____

Generic Questions for Materials

Output Variables

☐ What output variables define quality of results?

☐ How are the output variables measured?

☐ What are the targets or optimal ranges for output variables?

☐ What is the impact of output materials on PEMEO?

☐ What is the impact of PEMEO on output materials?

Input Variables

☐ What are the input variables?

☐ Which input variables are changed by the process?

☐ Which input variables must not be changed?

☐ How are the input variables measured?

☐ What are the targets or optimal ranges for input variables?

☐ Which input variables are static? dynamic?

☐ Which input variables are controllable? non-controllable?

☐ What is the impact of input materials on PEMEO?

☐ What is the impact of PEMEO on input materials?

Process Variables

☐ Which process variables change the materials?

☐ How are the process variables that change the materials measured?

☐ Which process variables that change the materials are static? dynamic?

☐ Which process variables that change the materials are controllable? non-controllable?

☐ Which process variables can downgrade the results?

☐ What is the impact of process on PEMEO?

☐ What is the impact of PEMEO on process?

Change in Variables

☐ What are the indications that the process variable has changed?

☐ What are the potential consequences if a process variable changes?

☐ What do I do if the process variables change?

☐ What changes do I have to make to the process variables if the input variables change?

☐ What changes to the input or process variables do I have to make if the quality standards for results change?

Another book by Gordon D. Shand

Interviewing to Gather Relevant Content for Training

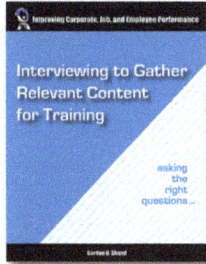

Effective training contributes to business success—
improved corporate, job, and employee performance.
But how do you figure out what training is effective?
This book provides the strategies you need to identify
training that will give you the best return on your
investment in training.

Part A:
• provides criteria and strategies you can use to identify:
  – training content that is relevant
  – what content you should address and not address
• describes pitfalls that you can encounter and ways to resolve these
  pitfalls

Part B describes an interviewing process where you provide leadership to
identify and gather content that is relevant, useful, and practical. You will
learn how to:
• help the subject matter expert provide quality content
• select content that is relevant and eliminate content that will not
  improve performance
• keep the subject matter expert engaged
• structure the content to effectively and efficiently develop training and
  assessment resources

The suggestions in this book are the accumulated experiences of many
training and performance consultants who have encountered the
challenges of gathering relevant content and developing effective training.

Who can benefit?

• educational, training, and performance consultants
• training program designers
• instructional designers
• technical writers
• trainers and coaches
• internal staff who develop training

www.ingramcontent.com/pod-product-compliance
Lightning Source LLC
Chambersburg PA
CBHW050240220326
41598CB00047B/7466